AS-Level
Chemistry

The Revision Guide

Exam Board: AQA

Editors:
Mary Falkner, Sarah Hilton, Paul Jordin, Sharon Keeley, Simon Little, Andy Park.

Contributors:
Vikki Cunningham, Ian H. Davis, John Duffy, Max Fishel, Emma Grimwood,
Richard Harwood, Lucy Muncaster, Derek Swain, Paul Warren, Chris Workman.

Proofreaders:
Barrie Crowther, Glenn Rogers, Julie Wakeling.

Published by Coordination Group Publications Ltd.

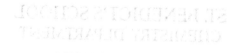

ISBN: 978 1 84762 123 8

With thanks to Jan Greenway for the copyright research.

With thanks to Science Photo Library for permission to reproduce the photograph used on page 62.

Groovy website: www.cgpbooks.co.uk
Jolly bits of clipart from CorelDRAW®
Printed by Elanders Hindson Ltd, Newcastle upon Tyne.

Contents

How Science Works

The Scientific Process 2

Unit 1: Section 1 — Atomic Structure

The Atom 4
Atomic Models 6
Relative Mass 8
Electronic Structure 10
Ionisation Energies 12

Unit 1: Section 2 — Amount of Substance

The Mole 14
Equations and Calculations 16
Titrations 18
Formulas, Yield and Atom Economy 20

Unit 1: Section 3 — Bonding and Periodicity

Ionic Bonding 22
Covalent Bonding 24
Shapes of Molecules 26
Polarisation and Intermolecular Forces 28
Metallic Bonding and Properties of Structures .. 31
Periodicity 34

Unit 1: Section 4 — Alkanes and Organic Chemistry

Basic Stuff 36
Formulas and Structural Isomerism 38
Alkanes and Petroleum 40
Alkanes as Fuels 42

Unit 2: Section 1 — Energetics

Enthalpy Changes 44
Calculating Enthalpy Changes 46

Unit 2: Section 2 — Kinetics and Equilibria

Reaction Rates and Catalysts 48
Reversible Reactions 51

Unit 2: Section 3 — Reactions and Elements

Redox Reactions 54
Group 7 — The Halogens 56
Halide Ions 58
Group 2 — The Alkaline Earth Metals 60
Extraction of Metals 63

Unit 2: Section 4 — More Organic Chemistry

Synthesis of Chloroalkanes 66
Haloalkanes 68
Reactions of Alkenes 71
E/Z Isomers and Polymers 74
Alcohols 76
Oxidising Alcohols 78
Analytical Techniques 80

Practical and Investigative Skills 83

Answers 86

Index 92

The Scientific Process

'How Science Works' is all about the scientific process — how we develop and test scientific ideas.
It's what scientists do all day, every day (well except at coffee time — never come between scientists and their coffee).

Scientists Come Up with **Theories** — Then **Test Them**...

Science tries to explain **how** and **why** things happen. It's all about seeking and gaining **knowledge** about the world around us. Scientists do this by **asking** questions and **suggesting** answers and then **testing** them, to see if they're correct — this is the **scientific process**.

1) **Ask** a question — make an **observation** and ask **why or how** whatever you've observed happens.
   E.g. Why does sodium chloride dissolve in water?

2) **Suggest** an answer, or part of an answer, by forming a **theory** or a **model** (a possible **explanation** of the observations or a description of what you think is happening actually happening).
   E.g. Sodium chloride is made up of charged particles which are pulled apart by the polar water molecules.

3) Make a **prediction** or **hypothesis** — a **specific testable statement**, based on the theory, about what will happen in a test situation.
   E.g. A solution of sodium chloride will conduct electricity much better than water does.

4) Carry out **tests** — to provide **evidence** that will support the prediction or refute it.
   E.g. Measure the conductivity of water and of sodium chloride solution.

The evidence supported Quentin's Theory of Flammable Burps.

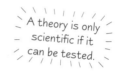

A theory is only scientific if it can be tested.

...Then They **Tell** Everyone About Their **Results**...

The results are **published** — scientists need to let others know about their work. Scientists publish their results in **scientific journals**. These are just like normal magazines, only they contain **scientific reports** (called papers) instead of the latest celebrity gossip.

1) Scientific reports are similar to the **lab write-ups** you do in school. And just as a lab write-up is **reviewed** (marked) by your teacher, reports in scientific journals undergo **peer review** before they're published.

   Scientists use standard terminology when writing their reports. This way they know that other scientists will understand them. For instance, there are internationally agreed rules for naming organic compounds, so that scientists across the world will know exactly what substance is being referred to. See page 36.

2) The report is sent out to **peers** — other scientists who are experts in the **same area**. They go through it bit by bit, examining the methods and data, and checking it's all clear and logical. When the report is approved, it's **published**. This makes sure that work published in scientific journals is of a **good standard**.

3) But peer review **can't guarantee** the science is **correct** — other scientists still need to **reproduce** it.

4) Sometimes **mistakes** are made and bad work is published. Peer review **isn't perfect** but it's probably the best way for scientists to self-regulate their work and to publish **quality reports**.

...Then **Other Scientists** Will **Test** the Theory Too

1) Other scientists read the published theories and results, and try to **test the theory** themselves. This involves:
   • Repeating the **exact same experiments**.
   • Using the theory to make **new predictions** and then testing them with **new experiments**.

2) If all the experiments in the world provide evidence to back it up, the theory is thought of as **scientific 'fact'** (for now).

3) If **new evidence** comes to light that **conflicts** with the current evidence the theory is questioned all over again. More rounds of **testing** will be carried out to try to find out where the theory **falls down**.

This is how the scientific process works — evidence supports a theory, loads of other scientists read it and test it for themselves, eventually all the scientists in the world agree with it and then bingo, you get to learn it.

This is exactly how scientists arrived at the structure of the atom (see pages 6-7) — and how they came to the conclusion that electrons are arranged in shells and orbitals (see page 10). It took years and years for these models to be developed and accepted — this is often the case with the scientific process.

The Scientific Process

If the **Evidence** Supports a Theory, It's **Accepted** — for Now

Our currently accepted theories have survived this '**trial by evidence**'. They've been tested **over and over again** and each time the results have backed them up. **BUT**, and this is a big but (teehee), they never become totally indisputable fact. Scientific **breakthroughs or advances** could provide new ways to question and test the theory, which could lead to **changes and challenges** to it. Then the testing starts all over again...

And this, my friend, is the **tentative nature of scientific knowledge** — it's always **changing** and **evolving**.

When CFCs were first used in fridges in the 1930s, scientists thought they were problem-free — well, why not? There was no evidence to say otherwise. It was decades before anyone found out that CFCs were actually making a whopping great hole in the ozone layer. See page 67.

Evidence Comes From **Lab Experiments**...

1) Results from **controlled experiments** in **laboratories** are **great**.
2) A lab is the easiest place to **control variables** so that they're all **kept constant** (except for the one you're investigating).
3) This means you can draw meaningful **conclusions**.

For example, if you're investigating how temperature affects the rate of a reaction you need to keep everything but the temperature constant, e.g. the pH of the solution, the concentration of the solution, etc.

...But You **Can't** Always do a Lab Experiment

There are things you **can't** study in a lab. And outside the lab controlling the variables is tricky, if not impossible.

- *Are increasing CO_2 emissions causing climate change?*
  There are other variables which may have an effect, such as changes in solar activity. You can't easily rule out every possibility. Also, climate change is a very **gradual process**. Scientists won't be able to tell if their predictions are correct for donkey's years.

See pages 42-43 for more on climate change.

- *Does drinking chlorinated tap water increase the risk of developing certain cancers?*
  There are always differences between groups of people. The best you can do is to have a **well-designed study** using **matched groups** — **choose two groups** of people (those who drink tap water and those who don't) which are **as similar as possible** (same mix of ages, same mix of diets etc.). But you still can't rule out every possibility. Taking newborn identical twins and treating them identically, except for making one drink gallons of tap water and the other only pure water, might be a fairer test, but it would present huge **ethical problems**.

Samantha thought her study was very well designed — especially the fitted bookshelf.

Science Helps to Inform **Decision-Making**

Lots of scientific work eventually leads to **important discoveries** that **could** benefit humankind — but there are often **risks** attached (and almost always **financial costs**).

Society (that's you, me and everyone else) must weigh up the information in order to **make decisions** — about the way we live, what we eat, what we drive, and so on. Information is also be used by **politicians** to devise policies and laws.

- **Chlorine** is added to water in **small quantities** to disinfect it. Some studies link drinking chlorinated water with certain types of cancer (see page 57). But the risks from drinking water contaminated by nasty bacteria are far, far greater. There are other ways to get rid of bacteria in water, but they're heaps **more expensive**.
- Scientific advances mean that **non-polluting hydrogen-fuelled cars** can be made. They're better for the environment, but are really expensive. Also, it'd cost a fortune to adapt the existing filling stations to store hydrogen.
- Pharmaceutical drugs are really expensive to develop, and drug companies want to make money. So they put most of their efforts into developing drugs that they can sell for a good price. Society has to consider the **cost** of buying new drugs — the **NHS** can't afford the most expensive drugs without **sacrificing** something else.

So there you have it — how science works...

Hopefully these pages have given you a nice intro to how science works, e.g. what scientists do to provide you with 'facts'. You need to understand this, as you're expected to know how science works yourself — for the exam and for life.

The Atom

This stuff about atoms and elements should be ingrained in your brain from GCSE. You do need to know it perfectly though if you are to negotiate your way through the field of man-eating tigers which is AS Chemistry.

Atoms are made up of **Protons**, **Neutrons** and **Electrons**

All elements are made of **atoms**. Atoms are made up of 3 types of particle — **protons**, **neutrons** and **electrons**.

Electrons
1) Electrons have **−1** charge.
2) They whizz around the nucleus in **orbitals**. The orbitals take up most of the **volume** of the atom.

Nucleus
1) Most of the **mass** of the atom is concentrated in the nucleus.
2) The **diameter** of the nucleus is rather titchy compared to the whole atom.
3) The nucleus is where you find the **protons** and **neutrons**.

The mass and charge of these subatomic particles is **really small**, so **relative mass** and **relative charge** are used instead.

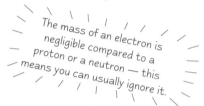
The mass of an electron is negligible compared to a proton or a neutron — this means you can usually ignore it.

Subatomic particle	Relative mass	Relative charge
Proton	1	+1
Neutron	1	0
Electron, e⁻	$\frac{1}{2000}$	−1

Nuclear Symbols Show Numbers of **Subatomic Particles**

You can figure out the **number** of protons, neutrons and electrons from the **nuclear symbol**.

Mass number
This tells you the **total** number of **protons** and **neutrons** in the nucleus.

Element symbol

$$_Z^A X$$

Atomic (proton) number
1) This is the number of **protons** in the nucleus — it identifies the element.
2) **All** atoms of the same element have the **same** number of protons.

Sometimes the atomic number is left out of the nuclear symbol, e.g. ⁷Li. You don't really need it because the element's symbol tells you its value.

1) For **neutral** atoms, which have no overall charge, the number of electrons is **the same as** the number of protons.
2) The number of neutrons is just **mass number minus atomic number**, i.e. 'top minus bottom' in the nuclear symbol.

Nuclear symbol	Atomic number, Z	Mass number, A	Protons	Electrons	Neutrons
$_3^7$ Li	3	7	3	3	7 − 3 = **4**
$_{35}^{80}$ Br	35	80	35	35	80 − 35 = **45**
$_{12}^{24}$ Mg	12	24	12	12	24 − 12 = **12**

"Hello, I'm Newt Ron..."

Ions have **Different** Numbers of **Protons and Electrons**

Negative ions have **more electrons** than protons...

E.g.

Br⁻   The negative charge means that there's 1 more electron than there are protons. Br has 35 protons (see table above), so Br⁻ must have 36 electrons. The overall charge = + 35 − 36 = −1.

...and **positive** ions have **fewer electrons** than protons. It kind of makes sense if you think about it.

E.g.

Mg²⁺   The 2+ charge means that there's 2 fewer electrons than there are protons. Mg has 12 protons (see table above), so Mg²⁺ must have 10 electrons. The overall charge = +12 − 10 = +2.

The Atom

Isotopes are Atoms of the Same Element with Different Numbers of Neutrons

Make sure you **learn** this definition and totally **understand** what it means —

Isotopes of an element are atoms with the same number of protons but different numbers of neutrons.

Chlorine-35 and chlorine-37 are examples of isotopes.

Different mass numbers mean different numbers of neutrons.

$35 - 17 = 18$ neutrons ← → $37 - 17 = 20$ neutrons

$$^{35}_{17}\text{Cl}$$

The **atomic numbers** are the same. **Both** isotopes have 17 protons and 17 electrons.

$$^{37}_{17}\text{Cl}$$

1) It's the **number** and **arrangement** of electrons that decides the **chemical properties** of an element. Isotopes have the **same configuration of electrons**, so they've got the **same** chemical properties.

2) Isotopes of an element do have slightly different **physical properties** though, such as different densities, rates of diffusion, etc. This is because **physical properties** tend to depend more on the **mass** of the atom.

Here's another example — naturally occurring **magnesium** consists of 3 isotopes.

^{24}Mg (79%)	^{25}Mg (10%)	^{26}Mg (11%)
12 protons	12 protons	12 protons
12 neutrons	**13** neutrons	**14** neutrons
12 electrons	12 electrons	12 electrons

The Periodic Table gives the atomic number for each element. The other number isn't the mass number, it's the relative atomic mass (see page 8). They're a bit different, but you can often assume they're equal — it doesn't matter unless you're doing really accurate work.

Practice Questions

Q1 Draw a diagram showing the structure of an atom, labelling each part.

Q2 Define the term 'isotope' and give an example.

Q3 Draw a table showing the relative charge and relative mass of the three subatomic particles found in atoms.

Q4 Using an example, explain the terms 'atomic number' and 'mass number'.

Q5 Where is the mass concentrated in an atom, and what makes up most of the volume of an atom?

Exam Questions

Q1 Hydrogen, deuterium and tritium are all isotopes of each other.
  a) Identify one similarity and one difference between these isotopes. [2 marks]
  b) Deuterium can be written as ^{2}H. Determine the number of protons, neutrons and electrons in a deuterium atom. [3 marks]
  c) Write the nuclear symbol for tritium, given that it has 2 neutrons. [1 mark]

Q2 This question relates to the atoms or ions A to D: A. $^{32}\text{S}^{2-}$, B. ^{40}Ar, C. ^{30}S, D. ^{42}Ca.
  a) Identify the similarity for each of the following pairs, justifying your answer in each case.
    (i)   A and B. [2 marks]
    (ii)   A and C. [2 marks]
    (iii) B and D. [2 marks]
  b) Which two of the atoms or ions are isotopes of each other? Explain your reasoning. [2 marks]

Got it learned yet? — Isotope so...

This is a nice straightforward page just to ease you in to things. Remember that positive ions have fewer electrons than protons, and negative ions have more electrons than protons. Get that straight in your mind or you'll end up in a right mess. There's nowt too hard about isotopes neither. They're just the same element with different numbers of neutrons.

Atomic Models

The model of the atom on the previous pages is darn useful for understanding loads of ideas in chemistry.
You can picture what's happening in your mind really well. But it is just a model. So it's not completely like that really.

The **Accepted Model** of the **Atom** Has **Changed** Throughout History

1) The model of the atom you're expected to know (the one on page 4) is one of the **currently accepted** ones. But in the past, **completely different** models were accepted, because they fitted the evidence available at the time.

2) As scientists did more experiments, **new evidence** was found and the models were **modified** to fit it.

3) At the start of the 19th century John Dalton described atoms as **solid spheres**, and said that different spheres made up the different elements.

delicious pudding

4) In 1897 **J J Thompson** concluded from his experiments that atoms **weren't** solid and indivisible. His measurements of **charge** and **mass** showed that an atom must contain even smaller, negatively charged particles — **electrons**. The 'solid sphere' idea of atomic structure had to be changed. The new model was known as the '**plum pudding model**'.

positively charged 'pudding'   electrons

Rutherford Showed that the **Plum Pudding** Model Was **Wrong**

1) In 1909 Ernest Rutherford and his students Hans Geiger and Ernest Marsden conducted the famous **gold foil experiment**. They fired **alpha particles** (which are positively charged) at an extremely thin sheet of gold.

2) From the plum pudding model, they were expecting **most** of the alpha particles to be deflected **very slightly** by the positive 'pudding' that made up most of an atom. In fact, most of the alpha particles passed **straight through** the gold atoms, and a very small number were deflected **backwards**. So the plum pudding model **couldn't be right**.

3) So Rutherford came up with a model that **could** explain this new evidence — the **nuclear model** of the atom. In this, there's a **tiny, positively charged nucleus** at the centre, surrounded by a '**cloud**' of **negative electrons** — most of the atom is **empty space**.

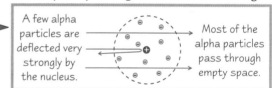

A few alpha particles are deflected very strongly by the nucleus.

Most of the alpha particles pass through empty space.

> This is nearly always the way scientific knowledge develops — **new evidence** prompts people to come up with **new, improved ideas**. Then other people go through each new, improved idea with a fine-tooth comb as well — modern '**peer review**' (see p2) is part of this process.

The **Refined Bohr Model** Explains a Lot...

1) There were quite a few other modifications to the model before we got to our currently accepted versions. Niels Bohr got pretty close though.

2) Scientists realised that electrons in a '**cloud**' around the nucleus of an atom, as Rutherford described, would quickly **spiral down** into the nucleus, causing the atom to **collapse**. Niels Bohr proposed a new model of the atom with four basic principles:

> 1) Electrons can only exist in **fixed orbits**, or **shells**, and not anywhere in between.
> 2) Each shell has a **fixed energy**.
> 3) When an electron moves between shells **electromagnetic radiation** is **emitted** or **absorbed**.
> 4) Because the energy of shells is fixed, the radiation will have a **fixed frequency**.

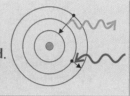

3) The frequencies of radiation emitted and absorbed by atoms were already known from experiments. The Bohr model **fitted these observations** — it looked good.

> One of the things that makes a theory **scientific** is that it's '**falsifiable**' — you can **make predictions** using the theory, then if you test the predictions and they turn out to be **wrong**, you know that the **theory's wrong**.

4) Scientists discovered that not all the electrons in a shell had the same energy. This meant that the Bohr model wasn't quite right. So, they **refined** it to include **subshells**.

Atomic Models

The **Bohr Model** Explained Why Some Gases are **Inert**

1) The Bohr model also explained why some elements (the noble gases) are **inert**.

2) Bohr said that the shells of an atom can only hold **fixed numbers of electrons**, and that an element's reactivity is due to its electrons. So, when an atom has **full shells** of electrons it's **stable** and does not react.

3) Loads of observations fitted in with the **Bohr model**, and the refined Bohr model was even better. But...

There's **More Than One** Model of Atomic Structure in Use Today

1) We now know that the refined Bohr model is **not perfect** — but it's still widely used to describe atoms because it is simple and explains many observations from experiments, like bonding and ionisation energy trends.

2) The most accurate model we have today involves complicated quantum mechanics. Basically, you can never know where an electron is or which direction it's going in at any moment, but you can say **how likely** it is to be at a certain point in the atom. Oh, and electrons can act as **waves** as well as particles. But you don't need to worry about that.

3) It might be **more accurate**, but it's a lot harder to get your head round and visualise. It **does** explain some observations that can't be accounted for by the Bohr model though.

4) So scientists use whichever model is most relevant to whatever they're investigating.

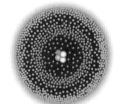

The quantum model of an atom with two shells of electrons. The denser the dots, the more likely an electron is to be there.

Practice Questions

Q1 Who developed the 'nuclear' model of the atom? What evidence did they have for it?

Q2 Describe the Bohr model of an atom. How was it later refined?

Q3 What predictions were made from the Bohr model that turned out to be correct?

Q4 Is there only one accepted model of the atom today?

Exam Questions

Q1 Read the passage below and then answer the questions that follow.

In 1911 Ernest Rutherford proposed a new model of the atom based on observations of the behaviour of atoms made by his students two years earlier. He said that atoms consisted of a small, positively charged nucleus around which negative electrons orbited. Scientists predicted that as electrons orbited the nucleus they would emit radiation with a continuous range of frequencies. This was tested and it was found that atoms emitted only certain fixed frequencies of radiation. In 1915 Niels Bohr proposed a model of the atom in which electrons were constrained to fixed orbits and could not exist anywhere between.

a) Why did Rutherford think that a new model of the atom was needed? [1 mark]

b) Why is Bohr's model thought to be a truer description of the atom than Rutherford's? [2 marks]

c) More accurate models of the atom have been developed since the Bohr model. Explain why the Bohr model is still used today. [1 mark]

Q2 The ion Ca^{2+} has the same electronic configuration as an argon atom, whilst the Ca^+ ion has an electronic configuration identical to that of a potassium atom.

a) Which of these two ions is most stable? [1 mark]

b) How do atomic models help to explain the relative stability of these ions? [2 marks]

These models are tiny — even smaller than size zero, I reckon...

The process of developing a model to fit the evidence available, looking for more evidence to show if it's correct or not, then revising the model if necessary is really important. It happens with all new scientific ideas. Remember, scientific 'facts' are only accepted as true because no one's proved yet that they aren't. It might all be bunkum.

Relative Mass

Relative mass...What? Eh?...Read on...

Relative Masses are Masses of Atoms Compared to Carbon-12

The actual mass of an atom is **very**, **very tiny**. Don't worry about exactly how tiny for now, but it's far **too small** to weigh. So, the mass of one atom is compared to the mass of a different atom. This is its **relative mass**. Here are some definitions to learn:

Relative atomic mass is an average, so it's not usually a whole number. Relative isotopic mass is always a whole number (at AS level anyway). E.g. a natural sample of chlorine contains a mixture of ^{35}Cl (75%) and ^{37}Cl (25%), so the relative isotopic masses are 35 and 37. But its relative atomic mass is 35.5.

The **relative atomic mass**, A_r, is the **average mass** of an atom of an element on a scale where an atom of **carbon-12** is 12.

Relative isotopic mass is the mass of an atom of an **isotope** of an element on a scale where an atom of **carbon-12** is 12.

The **relative molecular mass** (or **relative formula mass**), M_r, is the average mass of a **molecule** or **formula unit** on a scale where an atom of **carbon-12** is 12.

To find the relative molecular mass, just add up the relative atomic mass values of all the atoms in the molecule, e.g. $M_r(C_2H_6O) = (2 \times 12) + (6 \times 1) + 16 = 46$.

Relative formula mass is used for compounds that are ionic (or giant covalent, such as SiO_2). To find the relative formula mass, just add up the relative atomic masses (A_r) of all the ions in the formula unit. (A_r of ion = A_r of atom. The electrons make no difference to the mass.) E.g. $M_r(CaF_2) = 40 + (2 \times 19) = 78$.

Relative Masses can be Measured Using a Mass Spectrometer

You can use a **mass spectrometer** to find out loads of stuff. It can tell you the **relative atomic mass**, **relative molecular mass**, **relative isotopic abundance**, **molecular structure** and your **horoscope** for the next fortnight.

There are **5** things that happen when a sample is squirted into a mass spectrometer.

① **Vaporisation** — the sample is turned into **gas** (**vaporised**) using an electrical heater.

② **Ionisation** — the gas particles are bombarded with **high-energy electrons** to ionise them. Electrons are knocked off the particles, leaving **positive ions**.

③ **Acceleration** — the positive ions are accelerated by an **electric field**.

④ **Deflection** — The positive ions' paths are altered with a **magnetic field**. **Lighter ions** have less momentum and are deflected **more** than heavier ions. For a given magnetic field, **only** ions with a particular **mass/charge ratio** make it to the detector.

⑤ **Detection** — the magnetic field strength is **slowly increased**. As this happens, different ions (ones with a lower mass/charge ratio) can reach the detector. A **mass spectrum** is produced.

A Mass Spectrum

The **y-axis** gives the **abundance of ions**, often as a percentage. For an element, the **height** of each peak gives the **relative isotopic abundance**, e.g. 75.5% are the ^{35}Cl isotope.

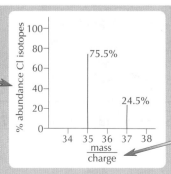

If the sample is an **element**, each line will represent a **different isotope** of the element.

The **x-axis** units are given as a '**mass/charge**' ratio. Since the charge on the ions is mostly **+1**, you can often assume the x-axis is simply the **relative isotopic mass**.

Relative Mass

A_r and *Relative Isotopic Abundance* can be Worked Out from a *Mass Spectrum*

You need to know how to calculate the **relative atomic mass** (A_r) of an element from the **mass spectrum**.

Here's how to calculate A_r for magnesium, using the mass spectrum below —

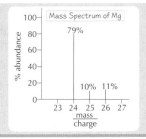

Step 1: For each peak, read the **% relative isotopic abundance** from the y-axis and the **relative isotopic mass** from the x-axis. **Multiply** them together to get the total mass for each isotope. $79 \times 24 = 1896$; $10 \times 25 = 250$; $11 \times 26 = 286$

Step 2: **Add** up these totals. $1896 + 250 + 286 = 2432$

Step 3: **Divide by 100** (since percentages were used). $A_r(Mg) = \dfrac{2432}{100} = 24.32 \approx \underline{\textbf{24.3}}$

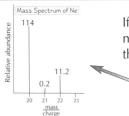

If the relative abundance is **not** given as a percentage, the total abundance may not add up to 100. In this case, don't panic. Just do steps 1 and 2 as above, but then divide by the **total relative abundance** instead of 100 — like this:

$$A_r(Ne) = \frac{(114 \times 20) + (0.2 \times 21) + (11.2 \times 22)}{114 + 0.2 + 11.2} \approx 20.18$$

Mass spectrometry is a good way to identify elements and molecules (it's kind of like fingerprinting).
For instance, small mass spectrometers have been used in probes to find out what the Martian atmosphere is made of.

Mass Spectrometry can be used to *Find Out* M_r

You can also get a mass spectrum for a **molecular sample**, such as ethanol (CH_3CH_2OH).

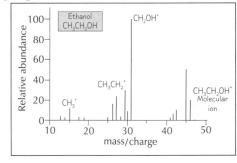

1) A **molecular ion**, $M^+_{(g)}$, is formed when the bombarding electrons remove 1 electron from the molecule. This gives the peak in the spectrum with the **highest mass** (furthest to the right, ignoring isotopes). The mass of M^+ gives **M_r** for the molecule, e.g. $CH_3CH_2OH^+$ has $M_r = 46$.

2) But it's not that simple — bombarding with electrons makes some molecules break up into fragments. These all show up on the mass spectrum, making a **fragmentation pattern**. For ethanol, the fragments you get include: CH_3^+ ($M_r = 15$), $CH_3CH_2^+$ ($M_r = 29$) and CH_2OH^+ ($M_r = 31$). Fragmentation patterns are actually pretty cool because you can use them to identify **molecules** and even their **structure**.
There's more about fragmentation patterns on p80.

Practice Questions

Q1 Explain what relative atomic mass (A_r) and relative isotopic mass mean.

Q2 Explain the difference between relative molecular mass and relative formula mass.

Q3 Describe how a mass spectrometer works.

Exam Questions

Q1 Copper, Cu, exists in two main isotopic forms, ^{63}Cu and ^{65}Cu.
   a) Calculate the relative atomic mass of Cu using the information from the mass spectrum.   [2 marks]
   b) Explain why the relative atomic mass of copper is not a whole number.   [2 marks]

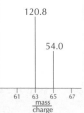

Q2 The percentage make-up of naturally occurring potassium is 93.11% ^{39}K, 0.12% ^{40}K and 6.77% ^{41}K.
   a) What method is used to determine the mass and abundance of each isotope?   [1 mark]
   b) Use the information to determine the relative atomic mass of potassium.   [2 marks]

You can't pick your relatives — you just have to learn them...

Working out M_r is dead easy — and using a calculator makes it even easier. It'll really help if you know the mass numbers for the first 20 elements or so, or you'll spend half your time looking back at the periodic table. I hope you've done the practice and exam questions, cos they pretty much cover the rest of the stuff, and if you can get them right, you've nailed it.

Electronic Structure

Those little electrons prancing about like mini bunnies decide what'll react with what — it's what chemistry's all about.

Electron Shells are Made Up of Sub-Shells and Orbitals

1) In the currently accepted model of the atom, electrons have **fixed energies**.
They move around the nucleus in certain regions of the atom called **shells** or **energy levels**.

2) Each shell is given a number called the **principal quantum number**.
The **further** a shell is from the nucleus, the **higher** its energy and the **larger** its principal quantum number.

3) This model helps to explain why electrons are **attracted** to the nucleus, but are not **drawn into it** and destroyed.

4) **Experiments** show that not all the electrons in a shell have exactly the same energy.
The **atomic model** explains this — shells are divided up into **sub-shells** that have slightly different energies.
The sub-shells have different numbers of **orbitals** which can each hold up to **2 electrons**.

This table shows the number of electrons that fit in each type of sub-shell.

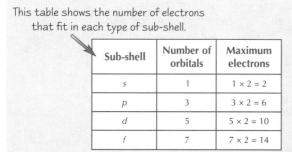

Sub-shell	Number of orbitals	Maximum electrons
s	1	1 × 2 = 2
p	3	3 × 2 = 6
d	5	5 × 2 = 10
f	7	7 × 2 = 14

And this one shows the sub-shells and electrons in the first four energy levels.

Shell	Sub-shells	Total number or electrons	
1st	1s	2	= 2
2nd	2s 2p	2 + (3 × 2)	= 8
3rd	3s 3p 3d	2 + (3 × 2) + (5 × 2)	= 18
4th	4s 4p 4d 4f	2 + (3 × 2) + (5 × 2) + (7 × 2)	= 32

5) The two electrons in each orbital spin in **opposite directions**.

Work Out Electron Configurations by Filling the Lowest Energy Levels First

You can figure out most electronic configurations pretty easily, so long as you know a few simple rules —

1) Electrons fill up the **lowest** energy sub-shells first.

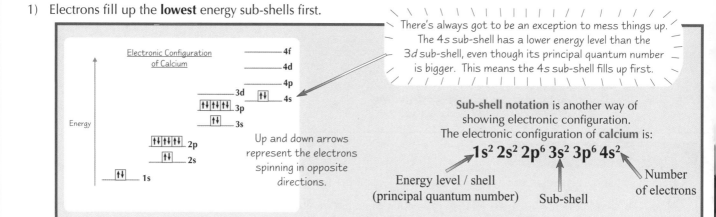

There's always got to be an exception to mess things up. The 4s sub-shell has a lower energy level than the 3d sub-shell, even though its principal quantum number is bigger. This means the 4s sub-shell fills up first.

Up and down arrows represent the electrons spinning in opposite directions.

Sub-shell notation is another way of showing electronic configuration.
The electronic configuration of **calcium** is:

$$1s^2\ 2s^2\ 2p^6\ 3s^2\ 3p^6\ 4s^2$$

Energy level / shell (principal quantum number)   Sub-shell   Number of electrons

2) Electrons fill orbitals **singly** before they start sharing.

	1s	2s	2p
Nitrogen	↑↓	↑↓	↑ ↑ ↑

	1s	2s	2p
Oxygen	↑↓	↑↓	↑↓ ↑ ↑

3) For the configuration of **ions** from the **s** and **p** blocks of the periodic table, just
remove or add the electrons to or from the highest energy occupied sub-shell.
E.g. $Mg^{2+} = 1s^2\ 2s^2\ 2p^6$, $Cl^- = 1s^2\ 2s^2\ 2p^6\ 3s^2\ 3p^6$

See the next page for more on the s and p block.

Watch out — **noble gas symbols**, like that of argon (Ar), are sometimes used in electron configurations.
For example, calcium ($1s^2\ 2s^2\ 2p^6\ 3s^2\ 3p^6\ 4s^2$) can be written as $[Ar]4s^2$, where $[Ar] = 1s^2\ 2s^2\ 2p^6\ 3s^2\ 3p^6$.

Electronic Structure

Transition Metals Behave Unusually

1) **Chromium** (Cr) and **copper** (Cu) are badly behaved. They donate one of their **4s** electrons to the **3d sub-shell**. It's because they're happier with a **more stable** full or half-full d sub-shell.

Cr atom (24 e⁻): $1s^2\ 2s^2\ 2p^6\ 3s^2\ 3p^6\ 3d^5\ 4s^1$ Cu atom (29 e⁻): $1s^2\ 2s^2\ 2p^6\ 3s^2\ 3p^6\ 3d^{10}\ 4s^1$

2) And here's another weird thing about transition metals — when they become **ions**, they lose their **4s** electrons **before** their 3d electrons.

Fe atom (26 e⁻): $1s^2\ 2s^2\ 2p^6\ 3s^2\ 3p^6\ 3d^6\ 4s^2$ → Fe^{3+} ion (23 e⁻): $1s^2\ 2s^2\ 2p^6\ 3s^2\ 3p^6\ 3d^5$

Electronic Structure Decides the Chemical Properties of an Element

The number of **outer shell electrons** decides the chemical properties of an element.

1) The **s block** elements (Groups 1 and 2) have 1 or 2 outer shell electrons. These are easily **lost** to form positive ions with an **inert gas configuration**. E.g. Na — $1s^2\ 2s^2\ 2p^6\ 3s^1$ → Na^+ — $1s^2\ 2s^2\ 2p^6$ (the electronic configuration of neon).

2) The elements in Groups 5, 6 and 7 (in the p block) can **gain** 1, 2 or 3 electrons to form negative ions with an **inert gas configuration**. E.g. O — $1s^2\ 2s^2\ 2p^4$ → O^{2-} — $1s^2\ 2s^2\ 2p^6$. Groups 4 to 7 can also **share** electrons when they form covalent bonds.

3) Group 0 (the inert gases) have **completely filled** s and p sub-shells and don't need to bother gaining, losing or sharing electrons — their full sub-shells make them **inert**.

4) The **d block elements** (transition metals) tend to **lose** s and d electrons to form positive ions.

Practice Questions

Q1 Write down the sub-shells in order of increasing energy up to 4p.

Q2 How many electrons would full s, p and d sub-shells contain?

Q3 Chromium and copper don't fill up their shells in the same way as other atoms. Explain the differences.

Q4 Which groups of the Periodic Table tend to gain electrons to form negative ions?

Exam Questions

Q1 Potassium reacts with oxygen to form potassium oxide, K_2O.

a) Give the electron configurations of the K atom and K^+ ion. [2 marks]

b) Using arrow-in-box notation, give the electron configuration of the oxygen atom. [2 marks]

c) Explain why it is the outer shell electrons, not those in the inner shells, which determine the chemistry of potassium and oxygen. [2 marks]

Q2 This question concerns the electron configurations in atoms and ions.

a) What is the electron configuration of a manganese atom? [1 mark]

b) Using arrow-in-box notation, give the electron configuration of the Al^{3+} ion. [2 marks]

c) Identify the element with the 4th shell configuration $4s^2 4p^2$. [1 mark]

d) Suggest the identity of an atom, a positive ion and a negative ion with the configuration $1s^2\ 2s^2\ 2p^6\ 3s^2\ 3p^6$. [3 marks]

She shells sub-sells on the shesore...

The way electrons fill up the orbitals is kind of like how strangers fill up seats on a bus. Everyone tends to sit in their own seat till they're forced to share. Except for the huge, scary, smelly man who comes and sits next to you. Make sure you learn the order the sub-shells are filled up, so you can write electron configurations for any atom or ion they throw at you.

Ionisation Energies

This page gets a trifle brain-boggling, so I hope you've got a few aspirin handy...

Ionisation is the Removal of One or More Electrons

When electrons have been removed from an atom or molecule, it's been **ionised**.
The energy you need to remove the first electron is called the **first ionisation energy** (or often just ionisation energy).

> The **first ionisation energy** is the energy needed to remove 1 electron from
> **each atom** in **1 mole** of **gaseous** atoms to form 1 mole of gaseous 1+ ions.

You can write **equations** for this process — here's the equation for the **first ionisation of oxygen** :

$$O_{(g)} \rightarrow O^+_{(g)} + e^- \quad \text{1st ionisation energy} = +1314 \text{ kJ mol}^{-1}$$

Here are a few rather important points about ionisation energies:

1) You **must** use the gas state symbol, **(g)**, because ionisation energies are measured for gaseous atoms.

2) Always refer to **1 mole** of atoms, as stated in the definition, rather than to a single atom.

3) The **lower** the ionisation energy, the **easier** it is to form an ion.

The Factors Affecting Ionisation Energy are...

Nuclear Charge — The **more protons** there are in the nucleus, the more positively charged the nucleus is and the **stronger the attraction** for the electrons.

Distance from Nucleus — Attraction falls off very **rapidly with distance**. An electron **close** to the nucleus will be **much more** strongly attracted than one further away.

Shielding — As the number of electrons **between** the outer electrons and the nucleus **increases**, the outer electrons feel less attraction towards the nuclear charge. This lessening of the pull of the nucleus by inner shells of electrons is called **shielding (or screening)**.

> A **high ionisation energy** means there's a **high attraction** between the **electron** and the **nucleus**.

Ionisation Energy Decreases Down Group 2

1) This provides **evidence** that electron shells **REALLY DO EXIST**.

2) If each element down Group 2 has an **extra electron shell** compared to the one above, the extra inner shells will **shield** the outer electrons from the attraction of the nucleus.

3) Also, the extra shell means that the outer electrons are **further away** from the nucleus, so the nucleus's attraction will be greatly reduced.

> It makes sense that both of these factors will make it **easier** to remove outer electrons, resulting in a **lower ionisation energy**.

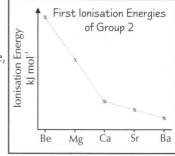

Ionisation Energy Increases Across a Period

The graph below shows the first ionisation energies of the elements in **Period 3**.

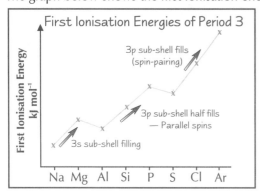

1) As you **move across** a period, the general trend is for the ionisation energies to **increase** — i.e. it gets harder to remove the outer electrons.

2) This can be explained because the number of protons is increasing, which means a stronger **nuclear attraction**.

3) All the extra electrons are at **roughly the same** energy level, even if the outer electrons are in different orbital types.

4) This means there's generally little **extra shielding** effect or **extra distance** to lessen the attraction from the nucleus.

5) But, there are **small drops** between Groups 2 and 3, and 5 and 6. Tell me more, I hear you cry. Well, alright then...

Ionisation Energies

The Drop between Groups 2 and 3 Shows **Sub-Shell Structure**

| Mg | $1s^2\,2s^2\,2p^6\,3s^2$ | 1st ionisation energy = 738 kJ mol^{-1} |
| Al | $1s^2\,2s^2\,2p^6\,3s^2\,3p^1$ | 1st ionisation energy = 578 kJ mol^{-1} |

1) Aluminium's outer electron is in a **3p orbital** rather than a 3s. The 3p orbital has a **slightly higher** energy than the 3s orbital, so the electron is, on average, to be found **further** from the nucleus.

2) The 3p orbital has additional shielding provided by the **3s^2 electrons**.

3) Both these factors together are strong enough to **override** the effect of the increased nuclear charge, resulting in the ionisation energy **dropping** slightly.

4) This pattern in ionisation energies provides **evidence** for the theory of electron sub-shells.

The Drop between Groups 5 and 6 is due to **Electron Repulsion**

| P | $1s^2\,2s^2\,2p^6\,3s^2\,3p^3$ | 1st ionisation energy = 1012 kJ mol^{-1} |
| S | $1s^2\,2s^2\,2p^6\,3s^2\,3p^4$ | 1st ionisation energy = 1000 kJ mol^{-1} |

1) The **shielding is identical** in the phosphorus and sulfur atoms, and the electron is being removed from an identical orbital.

2) In phosphorus's case, the electron is being removed from a **singly-occupied** orbital. But in sulfur, the **electron** is being **removed** from an orbital containing two electrons.

Phosphorus: (Ne) 3s [↑↓] 3p [↑][↑][↑]   Sulfur: (Ne) 3s [↑↓] 3p [↑↓][↑][↑]

The **repulsion** between two electrons in an orbital means that electrons are **easier to remove** from shared orbitals.

3) Yup, yet more **evidence** for the electronic structure model.

Practice Questions

Q1 Define first ionisation energy and give an equation as an example.

Q2 Describe the three main factors that affect ionisation energies.

Q3 When an atom is ionised, does it release or absorb energy?

Q4 Do ionisation energies increase or decrease as you go down Group 2?

Exam Questions

Q1 The first ionisation energies of the elements lithium to neon are given below in kJ mol^{-1}:

Li	Be	B	C	N	O	F	Ne
519	900	799	1090	1400	1310	1680	2080

a) Write an equation, including state symbols, to represent the first ionisation energy of lithium. [2 marks]

b) Explain why the ionisation energies show an overall tendency to increase across the period. [3 marks]

c) Explain the irregularities in this trend for:
(i) boron
(ii) oxygen [4 marks]

Q2 First ionisation energy decreases down Group 2.

Explain how this trend provides evidence for the arrangement of electrons in levels. [3 marks]

Shirt crumpled — ionise it...

When you're talking about ionisation energies in exams, always use the three main factors — shielding, nuclear charge and distance from nucleus. Make sure you understand how ionisation energies provide evidence that electron shells and subshells DO exist. They don't prove the model is right, but they do make the scientific community think it's a good'un.

The Mole

It'd be handy to be able to count out atoms — but they're way too tiny. You can't even see them, never mind get hold of them with tweezers. But not to worry — using the idea of relative mass, you can figure out how many atoms you've got.

A **Mole** is Just a (Very Large) **Number of Particles**

1) Amount of substance is measured using a unit called the **mole** (**mol** for short) and given the symbol **n**.
2) One mole is roughly **6×10^{23} particles** (**Avogadro's constant**).
3) It **doesn't matter** what the particles are. They can be atoms, molecules, electrons, ions, penguins — **anything**.

> In the reaction $C + O_2 \rightarrow CO_2$:
> **1 atom** of carbon reacts with **1 molecule** of oxygen to make **1 molecule** of carbon dioxide,
> so **1 mole** of carbon reacts with **1 mole** of oxygen to make **1 mole** of carbon dioxide.

> In the reaction $2Mg + O_2 \rightarrow MgO$:
> **2 moles** of magnesium react with **1 mole** of oxygen molecules to make **1 mole** of magnesium oxide.

Molar Mass is the Mass of **One Mole**

Molar mass, **M**, is the mass of **one mole** of something.

But the main thing to remember is:

> **Molar mass is just the same as the relative molecular mass, M_r**
> (or relative formula mass)

That's why the mole is such a ridiculous number of particles (6×10^{23}) — it's the number of particles for which the weight in g is the same as the relative molecular mass.

The only difference is you stick a 'g mol⁻¹' for grams per mole on the end...

> **Example:** Find the molar mass of $CaCO_3$.
> Relative formula mass, M_r, of $CaCO_3 = 40 + 12 + (3 \times 16) = 100$
> So the molar mass, M, is **100 g mol⁻¹** — i.e. 1 mole of $CaCO_3$ weighs 100 g.

Here's another formula. This one's really important — you need it **all the time**:

> $$\text{Number of moles} = \frac{\text{mass of substance}}{\text{molar mass}}$$

> **Example:** How many moles of aluminium oxide are present in 5.1 g of Al_2O_3?
>
> Molar mass of Al_2O_3 $= (2 \times 27) + (3 \times 16)$
> $= 102 \text{ g mol}^{-1}$
>
> Number of moles of Al_2O_3 $= \dfrac{5.1}{102} = $ **0.05 moles**

The **Concentration** of a Solution is Measured in **mol dm⁻³**

1) The **concentration** of a solution is how many **moles** are dissolved per **1 dm³** of solution. The units are **mol dm⁻³** (or M).
2) Here's the formula to find the **number of moles**.

1 dm³ = 1000 cm³ = 1 litre

> $$\text{Number of moles} = \frac{\text{Concentration} \times \text{Volume (in cm}^3)}{1000}$$

or just

> **Number of moles = Concentration × Volume (in dm³)**

> **Example:** What mass of sodium hydroxide needs to be dissolved in 50 cm³ of water to make a 2 M solution?
>
> $$\text{Number of moles} = \frac{2 \times 50}{1000} = 0.1 \text{ moles of NaOH}$$
>
> Molar mass, M, of NaOH = 23 + 16 + 1 = 40 g mol⁻¹
>
> Mass = number of moles × M = 0.1 × 40 = **4 g**

The Mole

All Gases Take Up the Same Volume under the Same Conditions

If temperature and pressure stay the same, **one mole** of **any** gas always has the **same volume**.
At **room temperature and pressure** (r.t.p.), this happens to be **24 dm³**, (r.t.p is 298 K (25 °C) and 100 kPa).
Here are two formulas for working out the number of moles in a volume of gas. Don't forget — **ONLY** use them for r.t.p.

$$\text{Number of moles} = \frac{\text{Volume in dm}^3}{24} \quad \text{OR} \quad \text{Number of moles} = \frac{\text{Volume in cm}^3}{24\,000}$$

Example: How many moles are there in 6 dm³ of oxygen gas at r.t.p.?

$$\text{Number of moles} = \frac{6}{24} = \textbf{0.25 moles of oxygen molecules}$$

Ideal Gas equation — pV = nRT

In the real world (and AQA exam questions), it's not always room temperature and pressure.
The **ideal gas equation** lets you find the **number of moles** in a certain volume at **any temperature and pressure**.

$$pV = nRT$$

Where:
p = pressure (Pa)
V = volume (m³)
n = number of moles
The gas constant. Don't worry about what it means. Just learn it. → $R = 8.31$ J K^{-1}mol^{-1}
T = temperature (K)

$1\ cm^3 = 1 \times 10^{-6}\ m^3$
$1\ dm^3 = 1 \times 10^{-3}\ m^3$

$K = °C + 273$

Example:

At a temperature of 60 °C and a pressure of 250 kPa, a gas occupied a volume of 1100 cm³ and had a mass of 1.6 g.
Find its relative molecular mass.

$$n = \frac{pV}{RT} = \left(\frac{(250 \times 10^3) \times (1.1 \times 10^{-3})}{8.31 \times 333} \right) = 0.1 \text{ moles}$$

$1100\ cm^3 = 1.1 \times 10^{-3}\ m^3$

If 0.1 moles is 1.6 g, then 1 mole = $\dfrac{1.6}{0.1} = 16$ g. So the relative molecular mass (M_r) is **16**.

Practice Questions

Q1 How many molecules are there in one mole of ethane molecules?

Q2 What volume does 1 mole of gas occupy at r.t.p.?

Q3 Write down the ideal gas equation.

Exam Questions

Q1 Calculate the mass of 0.36 moles of ethanoic acid, CH_3COOH. [2 marks]

Q2 What mass of H_2SO_4 is needed to produce 60 cm³ of 0.25 M solution? [2 marks]

Q3 What volume will be occupied by 88 g of propane gas (C_3H_8)
a) at r.t.p.? [2 marks]
b) at 35 °C and 100 kPa? [2 marks]

Put your back teeth on the scale and find out your molar mass...

You need this stuff for loads of the calculation questions you might get, so make sure you know it inside out. Before you start plugging numbers into formulas, make sure they're in the right units. If they're not, you need to know how to convert them or you'll be tossing marks out the window. Learn all the definitions and formulas, then have a bash at the questions.

Equations and Calculations

Balancing equations'll cause you a few palpitations — as soon as you make one bit right, the rest goes pear-shaped.

Balanced Equations have **Equal Numbers** of each Atom on **Both Sides**

1) Balanced equations have the **same number** of each atom on **both** sides. They're.. well... you know... balanced.

2) You can only add more atoms by adding **whole compounds**. You do this by putting a number **in front** of a compound or changing one that's already there. You **can't** mess with formulas — ever.

Example: Balance the equation $C_2H_6 + O_2 \rightarrow CO_2 + H_2O$.

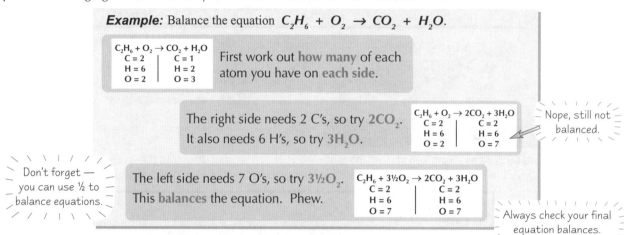

$C_2H_6 + O_2 \rightarrow CO_2 + H_2O$

C = 2	C = 1
H = 6	H = 2
O = 2	O = 3

First work out **how many** of each atom you have on **each side**.

The right side needs 2 C's, so try **2CO₂**. It also needs 6 H's, so try **3H₂O**.

$C_2H_6 + O_2 \rightarrow 2CO_2 + 3H_2O$

C = 2	C = 2
H = 6	H = 6
O = 2	O = 7

Nope, still not balanced.

Don't forget — you can use ½ to balance equations.

The left side needs 7 O's, so try **3½O₂**. This **balances** the equation. Phew.

$C_2H_6 + 3½O_2 \rightarrow 2CO_2 + 3H_2O$

C = 2	C = 2
H = 6	H = 6
O = 7	O = 7

Always check your final equation balances.

In **Ionic Equations** the **Charges** must Balance too

In ionic equations, only the **reacting particles** are included. You don't have to worry about the rest of the stuff.

Example: Balance the ionic equation $Cr_2O_7^{2-} + H^+ + e^- \rightarrow Cr^{3+} + H_2O$.

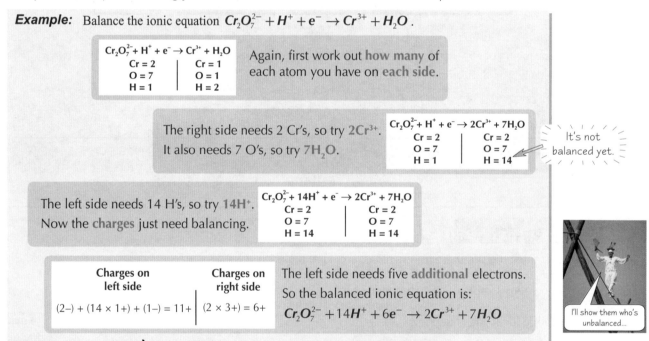

$Cr_2O_7^{2-} + H^+ + e^- \rightarrow Cr^{3+} + H_2O$

Cr = 2	Cr = 1
O = 7	O = 1
H = 1	H = 2

Again, first work out **how many** of each atom you have on **each side**.

The right side needs 2 Cr's, so try **2Cr³⁺**. It also needs 7 O's, so try **7H₂O**.

$Cr_2O_7^{2-} + H^+ + e^- \rightarrow 2Cr^{3+} + 7H_2O$

Cr = 2	Cr = 2
O = 7	O = 7
H = 1	H = 14

It's not balanced yet.

The left side needs 14 H's, so try **14H⁺**. Now the **charges** just need balancing.

$Cr_2O_7^{2-} + 14H^+ + e^- \rightarrow 2Cr^{3+} + 7H_2O$

Cr = 2	Cr = 2
O = 7	O = 7
H = 14	H = 14

Charges on left side	Charges on right side
$(2-) + (14 \times 1+) + (1-) = 11+$	$(2 \times 3+) = 6+$

The left side needs five **additional** electrons. So the balanced ionic equation is:

$$Cr_2O_7^{2-} + 14H^+ + 6e^- \rightarrow 2Cr^{3+} + 7H_2O$$

I'll show them who's unbalanced...

Balanced Equations can be used to Work out Masses

Example: Calculate the mass of iron oxide produced if 28 g of iron is burnt in air.

$$2Fe + \tfrac{3}{2}O_2 \rightarrow Fe_2O_3$$

The molar mass, M, of Fe = 56 g mol⁻¹, so the number of moles in 28 g of Fe $= \dfrac{mass}{M} = \dfrac{28}{56} = 0.5$ moles

From the equation: 2 moles of Fe produces 1 mole of Fe_2O_3, so 0.5 moles of Fe produces 0.25 moles of Fe_2O_3.

Once you know the number of moles and the molar mass (M) of Fe_2O_3, it's easy to work out the mass.

M of $Fe_2O_3 = (2 \times 56) + (3 \times 16) = 160$ g mol⁻¹

Mass of Fe_2O_3 = no. of moles × M = 0.25 × 160 = **40 g**. And that's your answer.

Equations and Calculations

That's not all... Balanced Equations can be used to Work Out Gas Volumes

It's pretty handy to be able to work out **how much gas** a reaction will produce, so that you can use **large enough apparatus**. Or else there might be a rather large bang.

Example: How much gas is produced when 15 g of sodium is reacted with excess water at r.t.p.?

$$2Na_{(s)} + 2H_2O_{(l)} \rightarrow 2NaOH_{(aq)} + H_{2(g)}$$

M of Na = 23 g mol^{-1}, so number of moles in 15 g of Na = $\frac{15}{23}$ = 0.65 moles

From the equation, 2 moles Na produces 1 mole H$_2$,

so you know 0.65 moles Na produces $\frac{0.65}{2}$ = 0.326 moles H$_2$.

So the volume of H$_2$ = 0.326 × 24 = **7.8 dm^3**

'Excess water' means you know all the sodium will react.

The reaction happens at room temperature and pressure, so you know 1 mole takes up 24 dm^3.

State Symbols Give a bit More Information about the Substances

State symbols are put after each compound in an equation. They tell you what **state of matter** things are in.

s = solid
l = liquid
g = gas
aq = aqueous
(solution in water)

To show you what I mean, here's an example —

$$CaCO_{3\,(s)} + 2HCl_{(aq)} \rightarrow CaCl_{2\,(aq)} + H_2O_{(l)} + CO_{2\,(g)}$$

solid   solution   solution   liquid   gas

Practice Questions

Q1 What is the state symbol for a solution of hydrochloric acid?

Q2 What is the difference between a full, balanced equation and an ionic equation?

Exam Questions

Q1 Calculate the mass of ethene required to produce 258 g of chloroethane, C$_2$H$_5$Cl.
$$C_2H_4 + HCl \rightarrow C_2H_5Cl$$
[4 marks]

Q2 15 g of calcium carbonate is heated strongly so that it fully decomposes. $CaCO_{3(s)} \rightarrow CaO_{(s)} + CO_{2(g)}$

a) Calculate the mass of calcium oxide produced. [3 marks]

b) Calculate the volume of gas produced. [3 marks]

Q3 Balance this equation: $KI + Pb(NO_3)_2 \rightarrow PbI_2 + 2KNO_3$ [1 mark]

Don't get in a state about equations...

Balancing equations is really, really important to the whole of AS Chemistry, so hang in there, and make sure you can do it. You can ONLY calculate reacting masses and gas volumes if you've got a balanced equation to work from. I've said it once, and I'll say it again — practise, practise, practise...it's the only road to salvation (by the way, where is salvation anyway?).

Titrations

*Titrations are used to find out the **concentration** of acid or alkali solutions. You're likely to have to do a **titration** for your **Practical Skills Assessment**, and even if you don't, you still need to know how to use titration results in **calculations**.*

Titrations need to be done Accurately

1) **Titrations** allow you to find out **exactly** how much acid is needed to **neutralise** a quantity of alkali.

2) You measure out some **alkali** using a pipette and put it in a flask, along with some **indicator**, e.g. **phenolphthalein**.

3) First of all, do a rough titration to get an idea where the **end point** is (the point where the alkali is **exactly neutralised** and the indicator changes colour). Add the **acid** to the alkali using a **burette** — giving the flask a regular **swirl**.

4) Now do an **accurate** titration. Run the acid in to within 2 cm³ of the end point, then add the acid **dropwise**. If you don't notice exactly when the solution changed colour you've **overshot** and your result won't be accurate.

5) **Record** the amount of acid used to **neutralise** the alkali. It's best to **repeat** this process a few times, making sure you get the same answer each time. This'll make sure your results are **reliable**.

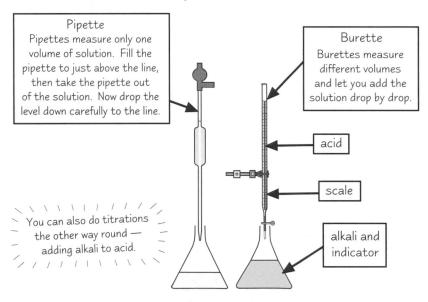

Pipette
Pipettes measure only one volume of solution. Fill the pipette to just above the line, then take the pipette out of the solution. Now drop the level down carefully to the line.

Burette
Burettes measure different volumes and let you add the solution drop by drop.

acid

scale

alkali and indicator

You can also do titrations the other way round — adding alkali to acid.

You can Calculate Concentrations from Titrations

Now for the calculations...

Example: 25 cm³ of 0.5 M HCl was used to neutralise 35 cm³ of NaOH solution. Calculate the concentration of the sodium hydroxide solution in mol dm⁻³.

First write a **balanced equation** and decide **what you know** and what you **need to know**:

$$HCl + NaOH \rightarrow NaCl + H_2O$$
$$25 \text{ cm}^3 \quad 35 \text{ cm}^3$$
$$0.5 \text{ M} \quad ?$$

It's just the formula from page 14.

Now work out how many **moles of HCl** you have:

$$\text{Number of moles HCl} = \frac{\text{concentration} \times \text{volume (cm}^3)}{1000} = \frac{0.5 \times 25}{1000} = 0.0125 \text{ moles}$$

From the equation, you know 1 mole of HCl neutralises 1 mole of NaOH.

So 0.0125 moles of HCl must neutralise **0.0125** moles of NaOH.

Now it's a doddle to work out the **concentration of NaOH**.

$$\text{Concentration of NaOH}_{(aq)} = \frac{\text{moles of NaOH} \times 1000}{\text{volume (cm}^3)} = \frac{0.0125 \times 1000}{35} = \textbf{0.36 mol dm}^{-3}$$

Titrations

You use a *Pretty Similar Method* to Calculate *Volumes* for Reactions

This is usually used for **planning experiments**.

You need to use this formula again, but this time **rearrange** it to find the volume. $\longrightarrow$

$$\text{number of moles} = \frac{\text{concentration} \times \text{volume (cm}^3)}{1000}$$

Example: 20.4 cm³ of a 0.5 M solution of sodium carbonate reacts with 1.5 M nitric acid. Calculate the volume of nitric acid required to neutralise the sodium carbonate.

Like before, first write a **balanced equation** for the reaction and decide **what you know** and what you **want to know**:

$$Na_2CO_3 + 2HNO_3 \rightarrow 2NaNO_3 + H_2O + CO_2$$

20.4 cm³ ?

0.5 M   1.5 M

Now work out how many **moles** of Na_2CO_3 you've got:

$$\text{No. of moles of } Na_2CO_3 = \frac{\text{concentration} \times \text{volume (cm}^3)}{1000} = \frac{0.5 \times 20.4}{1000} = 0.0102 \text{ moles}$$

1 mole of Na_2CO_3 neutralises 2 moles of HNO_3, so 0.0102 moles of Na_2CO_3 neutralises **0.0204 moles of HNO_3**.

Now you know the number of moles of HNO_3 and the concentration, you can work out the **volume**:

$$\text{Volume of } HNO_3 = \frac{\text{number of moles} \times 1000}{\text{concentration}} = \frac{0.0204 \times 1000}{1.5} = \textbf{13.6 cm}^3$$

Practice Questions

Q1 Explain what a titration is.

Q2 Write down the formula for calculating number of moles from the concentration and volume of a solution.

Q3 Rearrange this formula so that you could use it to calculate concentration. Then do the same for volume.

Exam Questions

Q1 Calculate the concentration in mol dm^{-3} of a solution of ethanoic acid, CH_3COOH, if 25.4 cm³ of it is neutralised by 14.6 cm³ of 0.5 M sodium hydroxide solution. $CH_3COOH + NaOH \rightarrow CH_3COONa + H_2O$ [3 marks]

Q2 You are supplied with 0.75 g of calcium carbonate and a solution of 0.25 M sulfuric acid. What volume of acid will be needed to neutralise the calcium carbonate? $CaCO_3 + H_2SO_4 \rightarrow CaSO_4 + H_2O + CO_2$ [4 marks]

Burettes and pipettes — big glass things, just waiting to be dropped...

Titrations are fiddly. But you do get to use big, impressive-looking equipment and feel like you're doing something important. Of course, then there's the results to do calculations with. The best way to start is always to write out the balanced equation and put what you know about each substance underneath it. Then think about what you're trying to find out.

Formulas, Yield and Atom Economy

Here's another page piled high with numbers — it's all just glorified maths really.

Empirical and Molecular Formulas are Ratios

You have to know what's what with empirical and molecular formulas, so here goes...

1) The **empirical formula** gives just the smallest whole number ratio of atoms in a compound.

2) The **molecular formula** gives the **actual** numbers of atoms in a molecule.

3) The molecular formula is made up of a whole **number** of empirical units.

> **Example:** A molecule has an empirical formula of $C_4H_3O_2$, and a molecular mass of 166 g. Work out its molecular formula.
>
> First find the empirical mass — $(4 \times 12) + (3 \times 1) + (2 \times 16)$
> $= 48 + 3 + 32 = 83$ g
>
> *Empirical mass is just like the relative formula mass... (if that helps at all...).*
>
> *Compare the empirical and molecular masses.*
>
> But the molecular mass is 166 g, so there are $\frac{166}{83} = 2$ empirical units in the molecule.
>
> The molecular formula must be the empirical formula × 2, so the molecular formula = $C_8H_6O_4$.

Empirical Formulas can be Calculated from Percentage Composition

You need to know how to work out empirical formulas from the **percentages** of the different elements.

> **Example:** A compound is found to have percentage composition 56.5% potassium, 8.7% carbon and 34.8% oxygen by mass. Calculate its empirical formula.
>
> *If you assume you've got 100 g of the compound, you can turn the % straight into mass, and then work out the number of moles as normal.*
>
> In **100 g** of compound there are:
>
> Use $n = \frac{mass}{M}$
>
> $\frac{56.5}{39} = 1.449$ moles of K $\quad \frac{8.7}{12} = 0.725$ moles of C $\quad \frac{34.8}{16} = 2.175$ moles of O
>
> Divide each number of moles by the **smallest number** — in this case it's 0.725.
>
> K: $\frac{1.449}{0.725} = 2.0$ $\quad$ C: $\frac{0.725}{0.725} = 1.0$ $\quad$ O: $\frac{2.175}{0.725} = 3.0$
>
> The ratio of K : C : O = 2 : 1 : 3. So you know the empirical formula's got to be K_2CO_3.

Percentage Yield Is Never 100%

1) The **theoretical yield** is the **mass of product** that **should** be formed in a chemical reaction. It assumes **no** chemicals are 'lost' in the process. You can use the **masses of reactants** and a **balanced equation** to calculate the theoretical yield for a reaction.

> **Example:** 1.40 g of iron filings is reacted with ammonia and sulfuric acid to make hydrated ammonium iron(II) sulfate.
>
> $Fe_{(s)} + 2NH_{3\,(aq)} + 2H_2SO_{4\,(aq)} + 6H_2O_{(l)} \rightarrow (NH_4)_2Fe(SO_4)_2.6H_2O_{(s)} + H_{2\,(g)}$
>
> Calculate the theoretical yield.
>
> Number of moles of **iron** ($A_r = 56$) reacted = mass ÷ molar mass = 1.40 ÷ 56 = **0.025 moles**.
>
> From the equation, 'moles of iron : moles of ammonium iron(II) sulfate' is 1:1, so 0.025 moles of product should form.
>
> Molar mass of $(NH_4)_2Fe(SO_4)_2.6H_2O_{(s)} = 392$, so **theoretical yield** = 0.025 × 392 = **9.8 g**.

2) For any reaction, the **actual** mass of product (the **actual yield**) will always be **less** than the theoretical yield. There are many reasons for this. For example, sometimes not all the 'starting' chemicals react fully. And some chemicals are always 'lost', e.g. some solution gets left on filter paper, or is lost during transfers between containers.

3) Once you've found the **theoretical yield** and the **actual yield**, you can work out the **percentage yield**. $\Longrightarrow$

$$\text{Percentage Yield} = \frac{\text{Actual Yield}}{\text{Theoretical Yield}} \times 100$$

4) So, in the ammonium iron(II) sulfate example above, the theoretical yield was 9.8 g. Say you weighed the hydrated ammonium iron(II) sulfate crystals produced and found the actual yield was **5.2 g**. Then

> **Percentage yield** = (5.2 ÷ 9.8) × 100 = **53%**

Formulas, Yield and Atom Economy

Atom Economy is a Measure of the Efficiency of a Reaction

1) The **efficiency** of a reaction is often measured by the **percentage yield**.
   This tells you how wasteful the **process** is — it's based on how much of the product is lost
   because of things like reactions not completing or losses during collection and purification.

2) But percentage yield doesn't measure how wasteful the **reaction** itself is. A reaction that has a 100%
   yield could still be very wasteful if a lot of the atoms from the **reactants** wind up in **by-products**
   rather than the **desired product**.

3) **Atom economy** is a measure of the proportion of reactant **atoms** that become part of the desired
   product (rather than by-products) in the **balanced** chemical equation.
   It's calculated using this formula:

$$\% \text{ atom economy} = \frac{\text{mass of desired product}}{\text{total mass of reactants}} \times 100$$

You can use the masses in grams, or their relative molecular masses.

Example: Bromomethane is reacted with sodium hydroxide to make methanol:

$$CH_3Br + NaOH \rightarrow CH_3OH + NaBr$$

Calculate the atom economy for this reaction.

Always make sure you're using a balanced equation.

$$\% \text{ atom economy} = \frac{\text{mass of desired product}}{\text{total mass of reactants}} \times 100$$

$$= \frac{(12+(3\times1)+16+1)}{(12+(3\times1)+80)+(23+16+1)} \times 100 = \frac{32\,g}{135\,g} \times 100 = \textbf{23.7\%}$$

The relative molecular masses have been used here. You need to use the numbers of moles from the balanced equation.

Practice Questions

Q1 Define 'empirical formula'.

Q2 What is the difference between a molecular formula and an empirical formula?

Q3 Give two examples of how chemicals could be 'lost' during a reaction.

Q4 Write down the formula for calculating percentage yield.

Q5 What is the difference between percentage yield and atom economy?

Exam Questions

Q1 Hydrocarbon X has a molecular mass of 78 g. It is found to have 92.3% carbon and 7.7%
hydrogen by mass. Calculate the empirical and molecular formulae of X. [3 marks]

Q2 Phosphorus trichloride (PCl_3) reacts with chlorine to give phosphorus pentachloride (PCl_5):

$$PCl_3 + Cl_2 \rightleftharpoons PCl_5$$

a) If 0.275 g of PCl_3 reacts with 0.142 g of chlorine, what is the theoretical yield of PCl_5? [2 marks]

b) When this reaction is performed 0.198 g of PCl_5 is collected. Calculate the percentage yield. [1 mark]

c) Changing conditions such as temperature and pressure will alter the percentage yield of this reaction.
Will changing these conditions affect the atom economy? Explain your answer. [2 marks]

The Empirical Strikes Back...

With this stuff, it's not enough to learn a few facts parrot-fashion, to regurgitate in the exam — you've gotta know how to use them. The only way to do that is to practise. Go through all the examples on these two pages again, this time working the answers out for yourself. Then test yourself on the practice exam questions. It'll help you sleep at night — honest.

Ionic Bonding

Every atom's aim in life is to have a full outer shell of electrons. Once they've managed this, that's it — they're happy.

Compounds are Atoms of Different Elements Bonded Together

1) When different elements join or bond together, you get a **compound**.
2) There are two main types of bonding in compounds — **ionic** and **covalent**. You need to make sure you've got them **both** totally sussed.

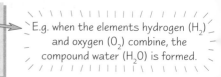
E.g. when the elements hydrogen (H_2) and oxygen (O_2) combine, the compound water (H_2O) is formed.

Ionic Bonding is when Ions are Stuck Together by Electrostatic Attraction

1) Ions are formed when electrons are **transferred** from one atom to another.
2) The simplest ions are single atoms which have either lost or gained 1, 2 or 3 electrons so that they've got a **full outer shell**. Here are some examples of ions:

A sodium atom (Na) **loses** 1 electron to form a sodium ion (Na^+) $Na \rightarrow Na^+ + e^-$
A magnesium atom (Mg) **loses** 2 electrons to form a magnesium ion (Mg^{2+}) $Mg \rightarrow Mg^{2+} + 2e^-$
A chlorine atom (Cl) **gains** 1 electron to form a chloride ion (Cl^-) $Cl + e^- \rightarrow Cl^-$
An oxygen atom (O) **gains** 2 electrons to form an oxide ion (O^{2-}) $O + 2e^- \rightarrow O^{2-}$

3) You **don't** have to remember what ion **each element** forms — nope, for many of them you just look at the Periodic Table. Elements in the same **group** all have the same number of **outer electrons**. So they have to **lose or gain** the same number to get the full outer shell that they're aiming for. And this means that they form ions with the **same charges**.
4) **Electrostatic attraction** holds positive and negative ions together — it's **very** strong. When atoms are held together like this, it's called **ionic bonding**.

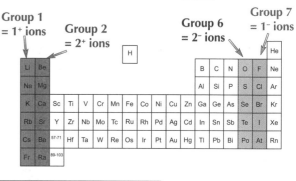

Sodium Chloride and Magnesium Oxide are Ionic Compounds

1) The formula of sodium chloride is **NaCl**. It just tells you that sodium chloride is made up of **Na^+ ions** and **Cl^- ions** (in a 1:1 ratio).
2) You can use '**dot-and-cross**' diagrams to show how ionic bonding works in sodium chloride —

Here, the dots represent the Na electrons and the crosses represent the Cl electrons (all electrons are really identical, but this is a good way of following their movement).

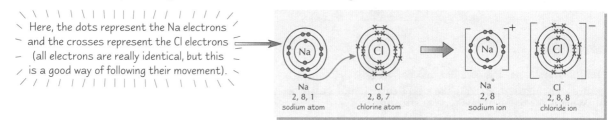

3) **Magnesium oxide**, MgO, is another good example:

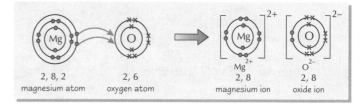

Dot (cross)

The positive charges in the compound **balance** the negative charges exactly — so the total overall charge is **zero**. This is a dead handy way of checking the formula.
- In **NaCl**, the single positive charge on the Na^+ ion balances the single negative charge on the Cl^- ion.
- In magnesium chloride, **$MgCl_2$**, the 2+ charge on the Mg^{2+} ion balances the two individual – charges on the two Cl^- ions.

Ionic Bonding

Sodium Chloride has a *Giant Ionic Lattice* Structure

1) Ionic crystals are giant lattices of ions. A **lattice** is just a **regular structure**.
2) The structure's called '**giant**' because it's made up of the same basic unit repeated over and over again.
3) In **sodium chloride**, the Na^+ and Cl^- ions are packed together. The sodium chloride lattice is **cube** shaped — different ionic compounds have different shaped structures, but they're all still giant lattices.

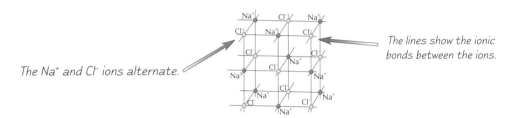

The Na^+ and Cl^- ions alternate.

The lines show the ionic bonds between the ions.

The structure of ionic compounds decides their **physical properties**...

Ionic Structure Explains the *Behaviour* of Ionic Compounds

1) **Ionic compounds conduct electricity when they're molten or dissolved — but not when they're solid.**
   The ions in a liquid are free to move (and they carry a charge).
   In a solid they're fixed in position by the strong ionic bonds.
2) **Ionic compounds have high melting points.**
   The giant ionic lattices are held together by strong electrostatic forces. It takes loads of energy to overcome these forces, so melting points are very high (801 °C for sodium chloride).
3) **Ionic compounds tend to dissolve in water.**
   Water molecules are polar — part of the molecule has a small negative charge, and the other bits have small positive charges (see p28). The water molecules pull the ions away from the lattice and cause it to dissolve.

Practice Questions

Q1 What's a compound?

Q2 Draw a dot-and-cross diagram showing the bonding between magnesium and oxygen.

Q3 What type of force holds ionic substances together?

Q4 Do ionic compounds tend to dissolve in water? Why?

Exam Questions

Q1 a) Draw a labelled diagram to show the structure of sodium chloride.   [3 marks]

   b) What is the name of this type of structure?   [1 mark]

   c) Would you expect sodium chloride to have a high or a low melting point?
       Explain your answer.   [4 marks]

Q2 a) Ions can be formed by electron transfer. Explain this and give an example
       of a positive and a negative ion.   [3 marks]

   b) Solid lead(II) bromide does not conduct electricity, but molten lead(II) bromide does.
       Explain this with reference to ionic bonding.   [3 marks]

Atom 1 says, "I think I lost an electron". Atom 2 replies, "are you positive?"...

Make sure that you can explain why ionic compounds do what they do. Their properties are all down to the fact that ionic crystals are made up of oppositely charged ions attracted to each other. Ionic bonding ONLY happens between a metal and a non-metal. If you've got two non-metal or two metals, they'll do different sorts of bonding — keep reading...

Covalent Bonding

And now for covalent bonding — this is when atoms share electrons with one another so they've all got full outer shells.

Molecules are Groups of Atoms Bonded Together

1) Molecules are the **smallest parts** of compounds that can take part in chemical reactions.

2) They're formed when **two or more** atoms bond together — it doesn't matter if the atoms are the **same** or **different**. Chlorine gas (Cl_2), carbon monoxide (CO), water (H_2O) and ethanol (C_2H_5OH) are all molecules.

3) Molecules are held together by strong **covalent bonds**.

In covalent bonding, two atoms **share** electrons, so they've **both got full outer shells** of electrons. Both the positive nuclei are attracted **electrostatically** to the shared electrons.

E.g. two iodine atoms bond covalently to form a molecule of iodine (I_2).

Covalent bonding happens between non-metals. Ionic bonding is between a metal and a non-metal.

Here's some more examples. These diagrams don't show all the electrons — just the ones in the **outer shells**:

   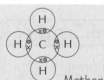

Hydrogen chloride, HCl — Hydrogen, H_2 — Water, H_2O — Methane, CH_4

There are Double and Triple Bonds Too

Atoms don't just form single bonds — **double** or even **triple covalent bonds** can form too.

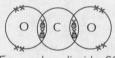

E.g. carbon dioxide, CO_2 — Nitrogen, N_2 (nitrogen's a triple-bonder)

It's not over yet — the **typical properties** of simple covalent molecules are covered on page 32.

There are Giant Covalent Structures Too

1) **Giant covalent** structures have a huge network of **covalently** bonded atoms. (They're sometimes called **macromolecular structures**.)

2) **Carbon** atoms can form this type of structure because they can each form **four** strong, covalent bonds. There are two types of giant covalent carbon structure you need to know about, **graphite** and **diamond**.

Graphite — Sheets of Hexagons with Delocalised Electrons

The carbon atoms are arranged in sheets of flat hexagons covalently bonded with three bonds each. The fourth outer electron of each carbon atom is delocalised.

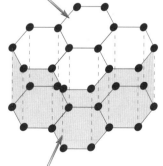

The sheets of hexagons are bonded together by weak van der Waals forces. (see p29)

Graphite's **structure** means it has **certain properties**:

1) The weak bonds **between** the layers in graphite are easily broken, so the sheets can slide over each other — graphite feels **slippery** and is used as a **dry lubricant** and in **pencils**.

2) The 'delocalised' electrons in graphite aren't attached to any particular carbon atoms and are **free to move** along the sheets, so an **electric current** can flow.

3) The layers are quite **far apart** compared to the length of the covalent bonds, so graphite has a **low density** and is used to make **strong, lightweight** sports equipment.

4) Because of the **strong covalent bonds** in the hexagon sheets, graphite has a **very high melting point** (it sublimes at over 3900 K).

'Sublimes' means it changes straight from a solid to a gas, skipping out the liquid stage.

5) Graphite is **insoluble** in any solvent. The covalent bonds in the sheets are **too difficult** to break.

Covalent Bonding

Diamond is the Hardest Known Substance

Diamond is also made up of **carbon atoms**. Each carbon atom is **covalently bonded** to **four** other carbon atoms. The atoms arrange themselves in a **tetrahedral** shape — its crystal lattice structure.

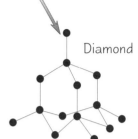
Diamond

Because of its **strong covalent** bonds:

1) Diamond has a **very high melting point** — it actually sublimes at over 3800 K.
2) Diamond is extremely **hard** — it's used in diamond-tipped drills and saws.
3) **Vibrations** travel easily through the stiff lattice, so it's a **good thermal conductor**.
4) It **can't conduct** electricity — all the outer electrons are held in localised bonds.
5) Like graphite, diamond won't dissolve in **any** solvent.

You can 'cut' diamond to form gemstones. Its structure
makes it refract light a lot, which is why it sparkles.

Dative Covalent Bonding is where Both Electrons come from One Atom

The **ammonium ion** (NH_4^+) is formed by dative covalent (or coordinate) bonding — it's an example the examiners love. It forms when the nitrogen atom in an ammonia molecule **donates a pair of electrons** to a proton (H^+) —

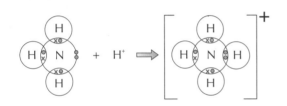
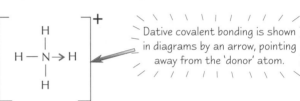

Dative covalent bonding is shown
in diagrams by an arrow, pointing
away from the 'donor' atom.

Practice Questions

Q1 Does covalent bonding occur between metal atoms or between non-metal atoms?

Q2 Describe how atoms are held together in covalent molecules.

Q3 Draw a dot-and-cross diagram to show the arrangement of the outer electrons in a molecule of iodine.

Q4 How are the carbon sheets in graphite held together?

Q5 In diamond, how many other carbons is each carbon atom bonded to?

Exam Questions

Q1 Methane, CH_4, is an organic molecule.
  a) What type of bonding would you expect it to have? [1 mark]
  b) Draw a dot-and-cross diagram to show the full electronic arrangement in a molecule of methane. [2 marks]

Q2 a) What type of bonding is present in the ammonium ion? [1 mark]
  b) Explain how this type of bonding occurs. [2 marks]

Q3 Carbon can be found as diamond and as graphite.
  a) What type of structure do diamond and graphite display? [1 mark]
  b) Draw diagrams to illustrate the structures of diamond and graphite. [2 marks]
  c) Compare and explain the electrical conductivities of diamond and
    graphite in terms of their structure and bonding. [4 marks]

Carbon is a girl's best friend...

More pretty diagrams to learn here folks — practise till you get every single dot and cross in the right place. It's totally amazing to think of these titchy little atoms sorting themselves out so they've got full outer shells of electrons. Remember — covalent bonding happens between two non-metals, whereas ionic bonding happens between a metal and a non-metal.

Shapes of Molecules

Chemistry would be heaps more simple if all molecules were flat. But they're not.

Molecular Shape depends on Electron Pairs around the Central Atom

Molecules and molecular ions come in loads of **different shapes**.

The shape depends on the **number of pairs** of electrons in the outer shell of the central atom.

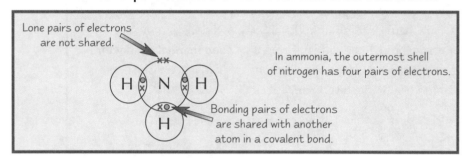

Lone pairs of electrons are not shared.

In ammonia, the outermost shell of nitrogen has four pairs of electrons.

Bonding pairs of electrons are shared with another atom in a covalent bond.

A lone pear

Electron Pairs exist as Charge Clouds

Bonding pairs and lone pairs of electrons exist as **charge clouds**.

A charge cloud is an area where you have a really **big chance** of finding an electron pair. The electrons don't stay still — they **whizz around** inside the charge cloud.

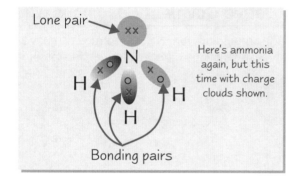

Lone pair

Here's ammonia again, but this time with charge clouds shown.

Bonding pairs

Electron Charge Clouds Repel Each Other

1) Electrons are all **negatively charged**, so it's pretty obvious that the charge clouds will **repel** each other as much as they can.

2) This sounds straightforward, but the **shape** of the charge cloud affects **how much** it repels other charge clouds. Lone-pair charge clouds repel **more** than bonding-pair charge clouds.

3) So, the **greatest** angles are between **lone pairs** of electrons, and bond angles between bonding pairs are often **reduced** because they are pushed together by lone-pair repulsion.

Lone-pair/lone-pair bond angles are the biggest.	*Lone-pair/bonding-pair bond angles are the second biggest.*	*Bonding-pair/bonding-pair bond angles are the smallest.*

4) This is known by the long-winded name '**Valence-Shell Electron-Pair Repulsion Theory**'.

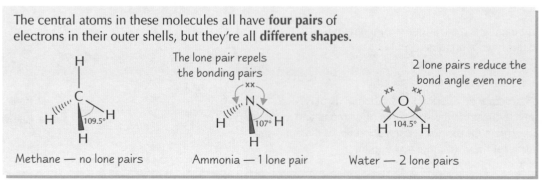

The central atoms in these molecules all have **four pairs** of electrons in their outer shells, but they're all **different shapes**.

The lone pair repels the bonding pairs

2 lone pairs reduce the bond angle even more

Methane — no lone pairs

Ammonia — 1 lone pair

Water — 2 lone pairs

In a molecule diagram, use wedges to show that a bond sticks out of the page towards you, and a broken (or dotted) line to show a bond goes behind the page.

5) These rules mean that the **shapes and bond angles** of loads of molecules can be predicted.

Shapes of Molecules

Practise **Drawing** these Molecules

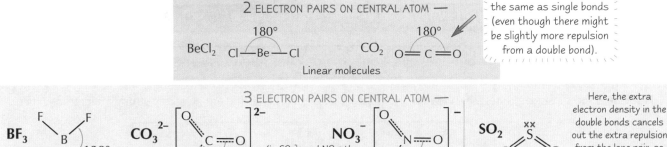

2 ELECTRON PAIRS ON CENTRAL ATOM —

Just treat double bonds the same as single bonds (even though there might be slightly more repulsion from a double bond).

$BeCl_2$ Cl—Be—Cl   180°

CO_2 O=C=O   180°

Linear molecules

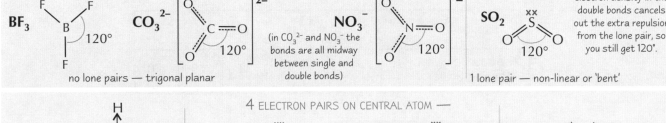

3 ELECTRON PAIRS ON CENTRAL ATOM —

BF_3 120°

no lone pairs — trigonal planar

CO_3^{2-} 120°

NO_3^- 120°
(in CO_3^{2-} and NO_3^- the bonds are all midway between single and double bonds)

SO_2 120°

1 lone pair — non-linear or 'bent'

Here, the extra electron density in the double bonds cancels out the extra repulsion from the lone pair, so you still get 120°.

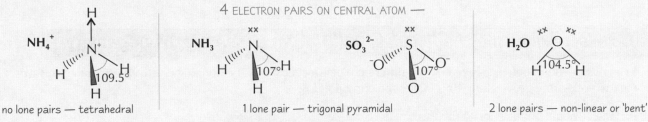

4 ELECTRON PAIRS ON CENTRAL ATOM —

NH_4^+ 109.5°

no lone pairs — tetrahedral

NH_3 107°

1 lone pair — trigonal pyramidal

SO_3^{2-} 107°

H_2O 104.5°

2 lone pairs — non-linear or 'bent'

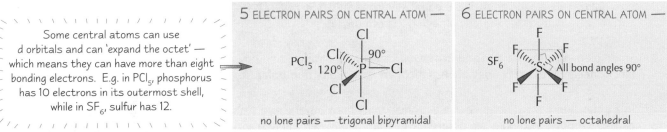

Some central atoms can use d orbitals and can 'expand the octet' — which means they can have more than eight bonding electrons. E.g. in PCl_5, phosphorus has 10 electrons in its outermost shell, while in SF_6, sulfur has 12.

5 ELECTRON PAIRS ON CENTRAL ATOM —

PCl_5 120°   90°

no lone pairs — trigonal bipyramidal

6 ELECTRON PAIRS ON CENTRAL ATOM —

SF_6 All bond angles 90°

no lone pairs — octahedral

Practice Questions

Q1 What is a lone pair of electrons?

Q2 What is a charge cloud?

Q3 Write down the order of the strength of repulsion between different kinds of electron pair.

Q4 Draw an example of a tetrahedral molecule.

Exam Question

Q1 Nitrogen and boron can form the chlorides NCl_3 and BCl_3.

   a) Draw dot-and-cross diagrams to show the bonding in NCl_3 and BCl_3.   [2 marks]

   b) Draw the shapes of the molecules NCl_3 and BCl_3.
       Show the approximate values of the bond angles on the diagrams and name each shape.   [6 marks]

   c) Explain why the shapes of NCl_3 and BCl_3 are different.   [3 marks]

These molecules ain't square...

In the exam, those evil examiners might try to throw you by asking you to predict the shape of an unfamiliar molecule. Don't panic — it'll be just like one you do know, e.g. PH_3 is the same shape as NH_3. Make sure you can draw every single molecule on this page. Yep, that's right — from memory. And learn what the shapes are called too.

Polarisation and Intermolecular Forces

Intermolecular forces hold molecules together. They're pretty important, cos we'd all be gassy clouds without them. Some of these intermolecular forces are down to polarisation. So you best make sure you know about that first...

Some Atoms **Attract** Bonding Electrons More than Other Atoms

The ability to attract the bonding electrons in a covalent bond is called electronegativity.

Fluorine is the most electronegative element. Oxygen, nitrogen and chlorine are also very strongly electronegative.

Element	H	C	N	Cl	O	F
Electronegativity (Pauling Scale)	2.1	2.5	3.0	3.0	3.5	4.0

Covalent Bonds may be Polarised by **Differences** in **Electronegativity**

In a covalent bond between two atoms of **different** electronegativities, the bonding electrons are **pulled towards** the more electronegative atom. This makes the bond **polar**.

1) The covalent bonds in diatomic gases (e.g. H₂, Cl₂) are **non-polar** because the atoms have **equal** electronegativities and so the electrons are equally attracted to both nuclei.

2) Some elements, like carbon and hydrogen, have pretty **similar** electronegativities, so bonds between them are essentially **non-polar**.

3) In a **polar bond**, the difference in electronegativity between the two atoms causes a **dipole**. A dipole is a **difference in charge** between the two atoms caused by a shift in **electron density** in the bond.

$\delta+$ $\delta-$
H —∘× Cl

'δ' (delta) means 'slightly', so '$\delta+$' means 'slightly positive'.

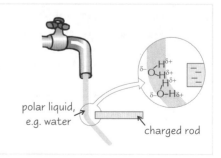

Permanent polar bonding

4) So what you need to **remember** is that the greater the **difference** in electronegativity, the **more polar** the bond.

Polar Molecules have Permanent Dipole-Dipole Forces

The $\delta+$ and $\delta-$ charges on **polar molecules** cause **weak electrostatic forces** of attraction **between** molecules.

E.g. hydrogen chloride gas has polar molecules.

 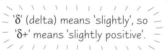

$\delta+$ $\delta-$ $\delta+$ $\delta-$ $\delta+$ $\delta-$
H—Cl ····· H—Cl ····· H—Cl

<u>Now this bit's pretty cool:</u>
If you put an **electrostatically charged rod** next to a jet of a polar liquid, like water, the liquid will **move** towards the rod. I wouldn't believe me either, but it's true. It's because **polar liquids** contain molecules with **permanent dipoles**. It doesn't matter if the rod is **positively** or **negatively** charged. The polar molecules in the liquid can **turn around** so the oppositely charged end is attracted towards the rod.

polar liquid, e.g. water

charged rod

Polarisation and Intermolecular Forces

Intermolecular Forces are **Very Weak**

Intermolecular forces are forces **between** molecules. They're much **weaker** than covalent, ionic or metallic bonds. There are three types you need to know about:

1) **Induced dipole-dipole** or **van der Waals** forces (this is the weakest type)
2) **Permanent dipole-dipole forces** (these are the ones that are caused by polar molecules — see the previous page)
3) **Hydrogen bonding** (this is the strongest type)

Van der Waals Forces are Found Between **All** Atoms and Molecules

Van der Waals forces cause **all** atoms and molecules to be **attracted** to each other.

1) **Electrons** in charge clouds are always **moving** really quickly. At any particular moment, the electrons in an atom are likely to be more to one side than the other. At this moment, the atom would have a **temporary dipole**.

2) This dipole can cause **another** temporary dipole in the opposite direction on a neighbouring atom. The two dipoles are then **attracted** to each other.

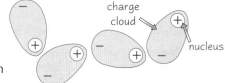

3) The second dipole can cause yet another dipole in a **third atom**. It's kind of like a domino rally.

4) Because the electrons are constantly moving, the dipoles are being **created** and **destroyed** all the time. Even though the dipoles keep changing, the **overall effect** is for the atoms to be **attracted** to each another.

Van der Waals Forces Can Hold Molecules in a **Lattice**

Van der Waals forces are responsible for holding **iodine** molecules together in a **lattice**.

1) Iodine atoms are held together in pairs by **strong** covalent bonds to form molecules of I_2.

2) But the molecules are then held together in a **molecular lattice** arrangement by **weak** van der Waals attractions.

Stronger **Van der Waals Forces** mean **Higher Boiling Points**

1) Not all van der Waals forces are the same strength — larger molecules have **larger electron clouds**, meaning **stronger** van der Waals forces.

2) Molecules with greater **surface areas** also have stronger van der Waals forces because they have a **more exposed electron cloud**.

3) When you **boil** a liquid, you need to **overcome** the intermolecular forces, so that the particles can **escape** from the liquid surface. It stands to reason that you need **more energy** to overcome **stronger** intermolecular forces, so liquids with stronger van der Waals forces will have **higher boiling points**.

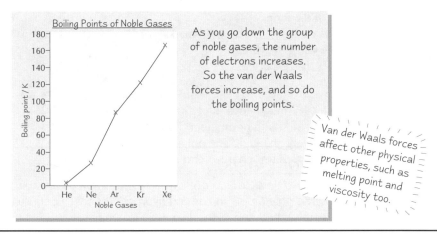

As you go down the group of noble gases, the number of electrons increases. So the van der Waals forces increase, and so do the boiling points.

Van der Waals forces affect other physical properties, such as melting point and viscosity too.

Polarisation and Intermolecular Forces

Hydrogen Bonding is the Strongest Intermolecular Force

1) Hydrogen bonding **only** happens when **hydrogen** is covalently bonded to **fluorine**, **nitrogen** or **oxygen**.

2) Fluorine, nitrogen and oxygen are very **electronegative**, so they draw the bonding electrons away from the hydrogen atom. The bond is so **polarised**, and hydrogen has such a **high charge density** because it's so small, that the hydrogen atoms form weak bonds with **lone pairs of electrons** on the fluorine, nitrogen or oxygen atoms of **other molecules**.

3) Molecules which have hydrogen bonding are usually **organic**, containing **-OH** or **-NH** groups. **Water** and **ammonia** both have hydrogen bonding.

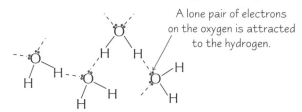

A lone pair of electrons on the oxygen is attracted to the hydrogen.

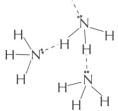

4) Hydrogen bonding has a **huge effect** on the properties of substances.

- Substances with hydrogen bonds have **higher boiling and melting points** than other similar molecules because of the **extra energy** needed to break the hydrogen bonds.

  This is the case with **water**, and also **hydrogen fluoride**, which has a much **higher boiling point** than the other hydrogen halides.

- Ice has more hydrogen bonds than liquid water, and hydrogen bonds are relatively **long**. So the H_2O molecules in ice are further apart on average, making ice **less dense** than liquid water.

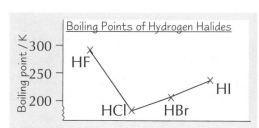

Practice Questions

Q1 What are the only bonds which can be purely non-polar?

Q2 What is the most electronegative element?

Q3 What is a dipole?

Q4 What's the strongest type of intermolecular force?

Q5 What is a hydrogen bond?

Exam Questions

Q1 Many covalent molecules have a permanent dipole, due to differences in electronegativities.
 a) Define the term electronegativity. [2 marks]
 b) Draw the shapes of the following molecules and mark any bond polarities clearly on your diagrams:
  (i) Br_2 (ii) H_2O (iii) NH_3 [5 marks]

Q2 a) Name three types of intermolecular force. [3 marks]
 b) Draw a clearly labelled diagram to show all the forms of intra- and intermolecular bonding in water. [4 marks]

Intra-molecular bonding is bonding inside molecules.

 c) This graph shows the boiling points of the Group 6 hydrides. Explain why water's boiling point is higher than expected in comparison to other Group 6 hydrides? [2 marks]

Van der Waal — a German hit for Oasis...

Just because intermolecular forces are a bit wimpy and weak, don't forget they're there. It'd all fall apart without them. Learn the three types — van der Waals, permanent dipole-dipole forces and hydrogen bonds. I bet fish are glad that water forms hydrogen bonds. If it didn't, their water would boil. (And they wouldn't have evolved in the first place.)

Metallic Bonding and Properties of Structures

Lots of this stuff you should already be able to recite in your sleep, but just in case it's fallen out of your brain, here it is...

Metals have Giant Structures

Metal elements exist as **giant metallic lattice structures**.

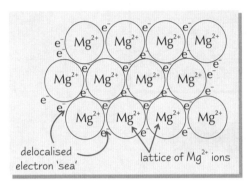

delocalised electron 'sea'

lattice of Mg^{2+} ions

1) The outermost shell of electrons of a metal atom is **delocalised** — the electrons are free to move about the metal. This leaves a **positive metal ion**, e.g. Na^+, Mg^{2+}, Al^{3+}.

2) The positive metal ions are **attracted** to the delocalised negative electrons. They form a lattice of closely packed positive ions in a **sea** of delocalised electrons — this is **metallic bonding**.

Metallic bonding explains why metals do what they do —

1) The **number of delocalised electrons per atom** affects the melting point. The **more** there are, the **stronger** the bonding will be and the **higher** the melting point. Mg^{2+} has **two** delocalised electrons per atom, so it's got a **higher melting point** than Na^+, which only has **one**. The **size** of the metal ion and the **lattice structure** also affect the melting point.

2) As there are **no bonds** holding specific ions together, the metal ions can slide over each another when the structure is pulled, so metals are **malleable** (can be shaped) and **ductile** (can be drawn into a wire).

3) The delocalised electrons can pass **kinetic energy** to each other, making metals **good thermal conductors**.

4) Metals are **good electrical conductors** because the **delocalised electrons** can carry a **current**.

5) Metals are **insoluble**, except in **liquid metals**, because of the **strength** of the metallic bonds.

The **Physical Properties** of Solids, Liquids and Gases Depend on **Particles**

1) A typical **solid** has its particles very **close** together. This gives it a high density and makes it **incompressible**. The particles **vibrate** about a **fixed point** and can't move about freely.

2) A typical **liquid** has a similar density to a solid and is virtually **incompressible**. The particles move about **freely** and **randomly** within the liquid, allowing it to flow.

3) In **gases**, the particles have **loads more** energy and are much **further apart**. So the density is generally pretty low and it's **very compressible**. The particles move about **freely**, with not a lot of attraction between them, so they'll quickly **diffuse** to fill a container.

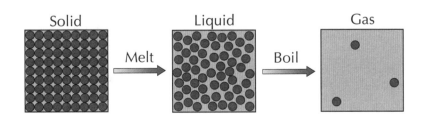

Solid   Melt →   Liquid   Boil →   Gas

The jelly state* occurs in solids when the particles start feeling a bit tired and achy.

*Don't write this in the exam, cos I just made it up, like...

Metallic Bonding and Properties of Structures

The Physical Properties of a **Solid** Depend on the **Nature** of its Particles

Here are some handy points that'll make AS chemistry a little less painful —

1) **Melting** and **boiling** points depend on **attraction** between particles.
2) The **closer** the particles, the **greater** the density.
3) If there are **charged** particles that are **free** to move, then it'll conduct electricity.
4) Solubility depends on the **type** of particles present.
   Water is a polar solvent and it tends to only dissolve other polar substances.
5) If a solid has a regular structure, it's called a **crystal**. The structure is a **crystal lattice**.

Covalent Bonds **Don't** Break during **Melting** and **Boiling***

This is something that confuses loads of people — prepare to be enlightened...

1) To **melt** or **boil** a simple covalent compound you only have to overcome the **van der Waals forces** or **hydrogen bonds** that hold the molecules together.
2) You **don't** need to break the much stronger covalent bonds that hold the atoms together in the molecules.
3) That's why simple covalent compounds have relatively **low melting** and **boiling points**. For example:

When you boil water, you don't get hydrogen and oxygen.

Chlorine, Cl_2, has **stronger** covalent bonds than bromine, Br_2.
But under normal conditions, chlorine is a **gas** and bromine a **liquid**.
Bromine has the higher boiling point because its molecules have **more electrons**, giving stronger van der Waals forces.

Except for giant molecular substances, like diamond.

Learn the **Properties** of the Main Substance Types

Make sure you know this stuff like the back of your spam —

Bonding	Examples	Melting and boiling points	Typical state at STP	Does solid conduct electricity?	Does liquid conduct electricity?	Is it soluble in water?
Ionic	NaCl $MgCl_2$	High	Solid	No (ions are held firmly in place)	Yes (ions are free to move)	Yes
Simple molecular (covalent)	CO_2 I_2 H_2O	Low (have to overcome van der Waals forces or hydrogen bonds, not covalent bonds)	May be solid (like I_2), but usually liquid or gas (water is liquid because it has hydrogen bonds)	No	No	Depends on how polarised the molecule is
Giant molecular (covalent)	Diamond Graphite SiO_2	High	Solid	No (except graphite)	— (will generally sublime)	No
Metallic	Fe Mg Al	High	Solid	Yes (delocalised electrons)	Yes (delocalised electrons)	No

Metallic Bonding and Properties of Structures

Bonding Models Match Observations

Scientists develop **models** based on **experimental evidence** — they're an attempt to **explain observations**. Bonding models explain how substances behave.

E.g. the **physical properties** of ionic compounds provide evidence that supports the theory of ionic bonding.
1) They have **high melting points** — this tells you that the atoms are held together by a **strong attraction**. Positive and negative ions are strongly attracted, so the **model** fits the **evidence**.
2) They are often **soluble** in **water** but **not** in **non-polar solvents** — this tells you that the particles are **charged**. The ions are **pulled apart** by **polar molecules** like water, but **not** by **non-polar** molecules. Again, the **model** of ionic structures fits this evidence.

Models of Bonding Have Their Limitations

Like pretty much all models, bonding models aren't totally accurate.
1) **Dot-and-cross models** of ionic and covalent bonding are great for explaining what's happening nice and clearly. But like most things in life, it's not really quite as simple as that.
2) One important reason is that most bonds aren't **purely ionic** or **purely covalent** but somewhere in between. This is down to **bond polarisation** (see page 28). Most compounds end up with a **mixture** of ionic and covalent properties.

Practice Questions

Q1 Why can metals conduct electricity?
Q2 Why are metals malleable?
Q3 Describe the motion of particles in solids, liquids and gases.
Q4 Why do gases diffuse to fill the space available?
Q5 What is a solid with a regular structure called?
Q6 What types of bonds must be overcome in order for a substance to boil or melt?
Q7 Do ionic compounds conduct electricity?

Exam Questions

Q1 Illustrate with a suitable labelled diagram the structure of calcium and explain what is meant by metallic bonding. [4 marks]

Q2

Substance	Melting point	Electrical conductivity of solid	Electrical conductivity of liquid	Solubility in water
A	High	Poor	Good	Soluble
B	Low	Poor	Poor	Insoluble
C	High	Good	Good	Insoluble
D	Very High	Poor	Poor	Insoluble

a) Identify the type of structure present in each substance, A to D. [4 marks]

b) Which substance is most likely to be:
(i) diamond, (ii) aluminium, (iii) sodium chloride and (iv) iodine? [2 marks]

Q3 Explain the electrical conductivity of magnesium, sodium chloride and graphite.
In your answer you should consider the structure and bonding of each of these materials. [12 marks]

Gases — like flies in jam jars...

You need to learn the info in the table on the left. With a quick glance in my crystal ball, I can almost guarantee you'll need a bit of it in your exam...let me look a bit closer and tell you which bit....mmm....nah. It's clouded over. You'll have to learn the lot. Sorry. Tell you what — close the book and see how much of the table you can scribble out from memory.

Periodicity

Periodicity is one of those words you hear a lot in Chemistry without ever really knowing what it means.
Well it basically means trends that occur (in physical and chemical properties) as you move across the periods.
E.g. Metal to non-metal is a trend that occurs going left to right in each period... The trends repeat each period.

The **Periodic Table** arranges Elements by **Proton Number**

1) The periodic table is arranged into **periods** (rows) and **groups** (columns), by atomic (proton) number.

2) All the elements **within a period** have the same number of **electron shells** (if you don't worry about s and p sub-shells) E.g. the elements in Period 2 have 2 electron shells.

3) All the elements **within a group** have the **same number** of electrons in their **outer shell** — so they have **similar properties**.

4) The **group number** tells you the number of electrons in the outer shell, e.g. Group 1 elements have 1 electron in their outer shell, Group 4 elements have 4 electrons and so on...

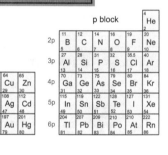

You can use the Periodic Table to work out **Electron Configurations**

The periodic table can be split into an **s block**, **d block** and **p block** like this: Doing this shows you which sub-shells all the electrons go into.

See page 10 if this sub-shell malarkey doesn't ring a bell.

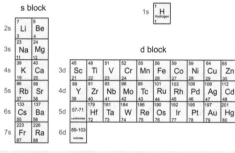

1) The **s-block** elements have an outer shell electron configuration of s¹ or s².

   Examples Lithium ($1s^2\ 2s^1$) and magnesium ($1s^2\ 2s^2\ 2p^6\ 3s^2$)

2) The **p-block** elements have an outer shell configuration of s^2p^1 to s^2p^6.

   Example Chlorine ($1s^2\ 2s^2\ 2p^6\ 3s^2\ 3p^5$)

3) The **d-block** elements have electron configurations in which d sub-shells are being filled.

   Example Cobalt ($1s^2\ 2s^2\ 2p^6\ 3s^2\ 3p^6\ 3d^7\ 4s^2$)

   Even though the 3d sub-shell fills last in cobalt, it's not written at the end of the line.

When you've got the periodic table **labelled** with the **shells** and **sub-shells** like the one up there, it's pretty easy to read off the electron structure of any element by starting at the top and working your way across and down until you get to your element.

A wee apology... This bit's really hard to explain clearly in words. If you're confused, just look at the examples until you get it...

Example

Electron structure of phosphorus (P):

Period 1 — $1s^2$ ⟵ Complete sub-shells
Period 2 — $2s^2\ 2p^6$ ⟵
Period 3 — $3s^2\ 3p^3$ ⟵ Incomplete outer sub-shell

So the full electron structure of phosphorus is: $1s^2\ 2s^2\ 2p^6\ 3s^2\ 3p^3$

Atomic Radius **Decreases** across a Period

1) As the number of protons increases, the **positive charge** of the nucleus increases. This means electrons are **pulled closer** to the nucleus, making the atomic radius smaller.

2) The extra electrons that the elements gain across a period are added to the **outer energy level** so they don't really provide any extra shielding effect (shielding works with inner shells mainly).

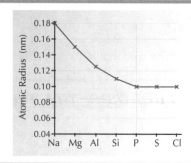

Periodicity

Melting and Boiling Points are linked to **Bond Strength** and **Structure**

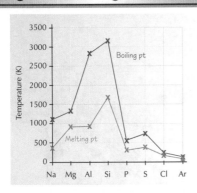

1) Sodium, magnesium and aluminium are **metals**. Their melting and boiling points **increase** across the period because the **metal-metal bonds** get stronger. The bonds get stronger because the metal ions have an increasing number of **delocalised electrons** and a decreasing **radius**. This leads to a higher **charge density**, which attracts the ions together more strongly.

2) Silicon is **macromolecular**, with a tetrahedral structure — **strong covalent bonds** link all its atoms together. **A lot** of energy is needed to break these bonds, so silicon has **high** melting and boiling points.

Sam is looking hot in the latest periodic trends.

3) Phosphorus (P_4), sulfur (S_8) and chlorine (Cl_2) are all **molecular substances**. Their melting and boiling points depend upon the strength of the **van der Waals forces** (see page 29) between their molecules. Van der Waals forces are weak and easily overcome so these elements have **low** melting and boiling points.

4) More atoms in a molecule mean stronger van der Waals forces. Sulfur is the **biggest molecule** (S_8), so it's got higher melting and boiling points than phosphorus or chlorine.

5) Argon has **very low** melting and boiling points because it exists as **individual atoms** (they're monatomic) resulting in **very weak** van der Waals forces.

Ionisation Energy Generally **Increases** across a Period

This is because of the **increasing attraction** between the outer shell electrons and the nucleus, due to the number of **protons** increasing (there are a few blips in the trend however — check back to pages 12-13 for more details).

Practice Questions

Q1 Which elements of Period 3 are found in the s block of the periodic table?

Q2 Write down the electronic configuration of sodium.

Q3 Which element in Period 3 has the largest atomic radius?

Q4 Which element in Period 3 has the highest melting point? Which has the highest boiling point?

Exam Questions

Q1 Explain why the melting point of magnesium is higher than that of sodium. [3 marks]

Q2 This table shows the melting points for the Period 3 elements.

Element	Na	Mg	Al	Si	P	S	Cl	Ar
Melting point / K	371	923	933	1680	317	392	172	84

In terms of structure and bonding explain why:

a) silicon has a high melting point.
b) the melting point of sulfur is higher than phosphorus.

Q3 State and explain the trend in atomic radius across Period 3. [4 marks]

Q4 Explain why the first ionisation energy of neon is greater than that of sodium. [2 marks]

Periodic trends — my mate Dom's always a decade behind...

*He still thinks Oasis, Blur and REM are the best bands around. The sad muppet. But not me. Oh no sirree, I'm up with the times — April Lavigne... Linkin' Pork... Christina Agorrilla. I'm hip, I'm with it. Da ga da ga da ga da ga.. ooaarrr ooup** * Obscure reference to Austin Powers: International Man of Mystery. You should watch it — it's better than doing Chemistry.*

Basic Stuff

There are zillions of organic compounds. And loads of them are pretty similar, but with a slight, but crucial difference. There's no way you could memorise all their names, so chemists have devised a clever way of naming each of them.

IUPAC Rules Help Avoid Confusion

The **IUPAC system** for naming organic compounds is the agreed **international language** of chemistry. Years ago, organic compounds were given whatever names people fancied, such as acetic acid and ethylene. But these names caused **confusion** between different countries.

The IUPAC system means scientific ideas can be communicated **across the globe** more effectively. So it's easier for scientists to get on with testing each other's work, and either confirm or dispute new theories.

Fancy a cup of tea? Tea, you know... Brown stuff. Oh come on, don't just sit there looking confused...

Nomenclature is a Fancy Word for Naming Organic Compounds

You can name **any** organic compound using **rules** of nomenclature.

Here's how the rules are used to name a **branched alkane** (there's more on alkanes on page 40).

1) Count the carbon atoms in the **longest continuous chain** — this gives you the stem:

Number of carbons	1	2	3	4	5	6
Stem	meth-	eth-	prop-	but-	pent-	hex-

Don't forget — the longest carbon chain may be bent.

2) Decide what **type** of molecule you've got. This gives you crucial parts of the name — see the table on the right.

The longest chain is **5** carbons, so the stem is **pent-**

It's a branched alkane, so the name's a bit more complicated than just **pentane**.

Homologous series	Prefix or Suffix	Example
alkanes	-ane	Propane $CH_3CH_2CH_3$
branched alkanes	-yl	methylpropane $CH_3CH(CH_3)CH_3$
alkenes	-ene	propene $CH_3CH=CH_2$
haloalkanes (halogenoalkanes)	chloro- bromo- iodo-	chloroethane CH_3CH_2Cl

3) Number the carbons in the **longest** carbon chain. If there's more than one longest chain, pick the one with the **most side-chains**.

With other compounds, you number so that the most important functional group (see p38) is on the lowest number carbon.

Longest chain with most side groups

4) Side-chains are added as prefixes at the start of the name. Put them in **alphabetical** order, with the **number** of the carbon atom each is attached to.

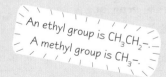

An ethyl group is CH_3CH_2-.
A methyl group is CH_3-.

5) If there's more than one **identical** side-chain or functional group, use **di-** (2), **tri-** (3) or **tetra-** (4) before that part of the name — but ignore this when working out the alphabetical order.

There's an ethyl group on carbon-3, and methyl groups on carbon-2 and carbon-4, so it's **3-ethyl-2,4-dimethylpentane**.

Basic Stuff

Name Haloalkanes Using the Same Rules

Haloalkanes are just alkanes where one or more hydrogens have been swapped for a **halogen**.

| dichloromethane | trichloromethane | 2-iodopropane | 2-bromo-2-chloro-1, 1, 1-trifluoroethane |

Naming Alkenes — Look at the Position of the Double Bond

Alkenes have at least one **double bond** in their carbon chain.
For alkenes with more than three carbons, you need to say which carbon the double bond starts from.

Example — 1) The longest chain is **5** carbons, so the stem of the name is **pent-**.

2) The functional group is **C=C**, so it's **pentene**.

3) Number the carbons from right to left (so the double bond starts on the lowest possible number). The first carbon in the double bond is **carbon 2**. So this is **pent-2-ene**.

Here are some more examples:

propene CH$_2$CHCH$_3$

Pent-2-ene CH$_3$CHCHCH$_2$CH$_3$

buta-1,3-diene CH$_2$CHCHCH$_2$

If the alkene has two double bonds the suffix becomes **diene**. The stem of the name usually gets an extra 'a' too (e.g. but<u>a</u>-, pent<u>a</u>- not but-, pent-) when there's more than one double bond. And you might see the numbers written first, e.g. 1,3-butadiene.

Practice Questions

Q1 What do you call an unbranched alkane with three carbon atoms?

Q2 In what order should prefixes be listed in the name of an organic compound?

Q3 What is a haloalkane?

Exam Questions

Q1 1-bromobutane is C$_4$H$_9$Br

a) Draw the structure of 1-bromobutane.   [1 mark]

b) Which homologous series does 1-bromobutane belong to?   [1 mark]

Homologous series are "families" of organic chemicals, see the table on the left.

c) 1-bromobutane can be made from the molecule shown on the right.
Name the molecule.   [2 marks]

Q2 Name the following molecules.

a)   [2 marks]

b)   [2 marks]

c)   [2 marks]

It's as easy as 1,2,3-trichloropent-2-ene...

The best thing to do now is find some random alkanes, alkenes and haloalkanes and work out their names using the rules.
Then have a bash at it the other way around — read the name and draw the compound. It might seem a wee bit tedious
now, but come the exam, you'll be thanking me. Doing the exam questions will give you some good practice too.

Formulas and Structural Isomerism

Isomers are fun — you can put the same atoms together in different ways to make completely different molecules. It's just like playing with plastic building bricks... But first some definitions to learn.

There are **Loads of Ways** of **Representing** Organic Compounds

TYPE OF FORMULA	WHAT IT SHOWS YOU	FORMULA FOR BUTANE
General formula	An algebraic formula that can describe **any member** of a family of compounds.	C_nH_{2n+2} (for all alkanes)
Empirical formula	The **simplest ratio** of atoms of each element in a compound (cancel the numbers down if possible). (So ethane, C_2H_6, has the empirical formula CH_3.)	C_2H_5
Molecular formula	The **actual** number of atoms of each element in a molecule, with any **functional groups** indicated.	C_4H_{10}
Structural formula	Shows the atoms **carbon by carbon**, with the attached hydrogens and functional groups.	$CH_3CH_2CH_2CH_3$
Displayed formula	Shows how all the atoms are **arranged**, and all the bonds between them.	

> A functional group is a reactive part of a molecule — it gives it most of its chemical properties. E.g. in an alcohol it's –OH.

A **homologous series** is a bunch of organic compounds which have the **same general formula**. Each member differs by $–CH_2–$. **Alkanes** are a homologous series.

Structural Isomers have different **Structural Arrangements** of Atoms

In structural isomers the atoms are **connected** in different ways. But they still have the **same molecular formula**. There are **three types** of structural isomers:

CHAIN ISOMERS

Chain isomers have different arrangements of the **carbon skeleton**. Some are **straight chains** and others **branched** in different ways.

butane   methylpropane

POSITIONAL ISOMERS

Positional isomers have the **same skeleton** and the **same atoms or groups of atoms** attached. The difference is that the atom or group of atoms is attached to a **different carbon atom**.

1-chlorobutane   2-chlorobutane

FUNCTIONAL GROUP ISOMERS

Functional group isomers have the same atoms arranged into **different functional groups**.

hex-1-ene   cyclohexane

Formulas and Structural Isomerism

Don't be Fooled — What Looks Like an Isomer Might Not Be

Atoms can rotate as much as they like around single **C–C bonds**. Remember this when you work out structural isomers — sometimes what looks like an isomer, isn't.

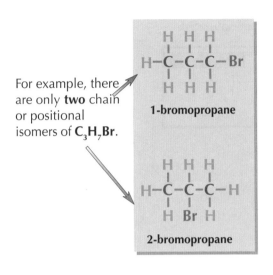

For example, there are only **two** chain or positional isomers of **C_3H_7Br**.

1-bromopropane

2-bromopropane

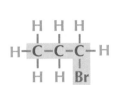

1-bromopropane again...

2-bromopropane again...

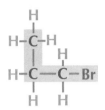

... and again
1-bromopropane

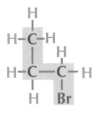

... and again
1-bromopropane

Practice Questions

Q1 Explain the difference between molecular formulas and structural formulas.

Q2 Draw the structural and displayed formulas of hexane.

Q3 What are structural isomers?

Q4 What is a positional isomer?

Exam Questions

Q1 There are four haloalkanes with the molecular formula C_4H_9Cl.

    a)   Give the names of all four of these haloalkanes.   [4 marks]

    b)   Identify a pair of positional isomers from your answer to part a).   [1 mark]

    c)   Identify a pair of chain isomers from your answer to part a).   [1 mark]

Q2 There are five chain isomers of the alkane C_6H_{14}.

    a)   Draw and name all five isomers of C_6H_{14}.   [10 marks]

    b)   Alkanes are an example of a homologous series. What is a homologous series?   [2 marks]

    c)   (i) Write down the molecular formula for an alkane molecule that has 8 carbon atoms.   [1 mark]

        (ii) Write out the full structural formula for the alkane molecule in part c)(i).
           Assume that it is an unbranched alkane.   [1 mark]

Q3 The alkane with the molecular formula C_5H_{12} has these chain isomers.

    a)   Name these isomers.   [3 marks]

    b)   Explain what is meant by the
        term 'chain isomerism'.   [2 marks]

Human structural isomers...

Alkanes and Petroleum

Alkanes are the first set of organic chemicals you need to know about. They're what petroleum's mainly made of.

Alkanes are **Saturated Hydrocarbons**

1) Alkanes have the **general formula C_nH_{2n+2}**. They've only got **carbon** and **hydrogen** atoms, so they're **hydrocarbons**.

2) Every carbon atom in an alkane has **four single bonds** with other atoms. It's **impossible** for carbon to make more than four bonds, so alkanes are **saturated**.

Here are a few examples of alkanes —

$$H-\underset{\underset{H}{|}}{\overset{\overset{H}{|}}{C}}-H \qquad H-\underset{\underset{H}{|}}{\overset{\overset{H}{|}}{C}}-\underset{\underset{H}{|}}{\overset{\overset{H}{|}}{C}}-H \qquad H-\underset{\underset{H}{|}}{\overset{\overset{H}{|}}{C}}-\underset{\underset{H}{|}}{\overset{\overset{H}{|}}{C}}-\underset{\underset{H}{|}}{\overset{\overset{H}{|}}{C}}-H$$

Methane   Ethane   Propane

3) You get **cycloalkanes** too. They have a ring of carbon atoms with two hydrogens attached to each carbon.

4) Cycloalkanes have a **different general formula** from that of normal alkanes (C_nH_{2n}, assuming they have only one ring), but they are still **saturated**.

Cyclohexane C_6H_{12}

cycloalkanes have two fewer hydrogens than other alkanes

Crude Oil is Mainly **Alkanes**

1) **Petroleum** is just a **poncy word** for crude oil — the black, yukky stuff they get out of the ground with huge oil wells. It's mostly **alkanes**. They range from **smallish alkanes**, like pentane, to **massive alkanes** with more than 50 carbons.

2) Crude oil isn't very useful as it is, but you can **separate** it into more useful bits (or **fractions**) by **fractional distillation**.

Here's how fractional distillation works — don't try this at home.

1) First, the crude oil is **vaporised** at about 350 °C.

2) The vaporised crude oil goes into the **fractionating column** and rises up through the trays. The largest hydrocarbons don't **vaporise** at all, because their boiling points are too high — they just run to the bottom and form a gooey **residue**.

3) As the crude oil vapour goes up the fractionating column, it gets **cooler**. Because of the different chain lengths, each fraction **condenses** at a different temperature. The fractions are **drawn off** at different levels in the column.

4) The hydrocarbons with the **lowest boiling points** don't condense. They're drawn off as **gases** at the top of the column.

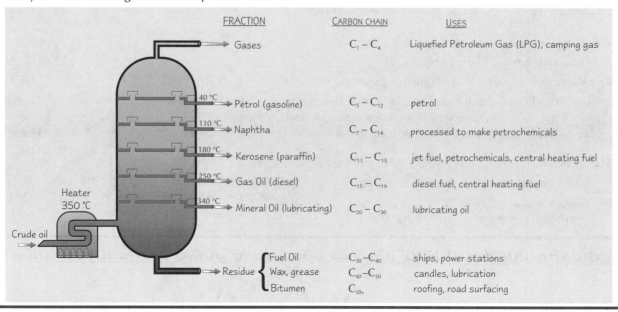

FRACTION	CARBON CHAIN	USES
Gases	$C_1 - C_4$	Liquefied Petroleum Gas (LPG), camping gas
Petrol (gasoline)	$C_5 - C_{12}$	petrol
Naphtha	$C_7 - C_{14}$	processed to make petrochemicals
Kerosene (paraffin)	$C_{11} - C_{15}$	jet fuel, petrochemicals, central heating fuel
Gas Oil (diesel)	$C_{15} - C_{19}$	diesel fuel, central heating fuel
Mineral Oil (lubricating)	$C_{20} - C_{30}$	lubricating oil
Fuel Oil	$C_{30} - C_{40}$	ships, power stations
Wax, grease	$C_{40} - C_{50}$	candles, lubrication
Bitumen	C_{50+}	roofing, road surfacing

40 °C, 110 °C, 180 °C, 250 °C, 340 °C

Heater 350 °C

Crude oil

Residue

Alkanes and Petroleum

Heavy Fractions can be 'Cracked' to Make Smaller Molecules

1) People want loads of the **light** fractions, like petrol and naphtha. They don't want so much of the **heavier** stuff like bitumen though. Stuff that's in high demand is much more **valuable** than the stuff that isn't.

2) To meet this demand, the less popular heavier fractions are **cracked**. Cracking is **breaking** long-chain alkanes into **smaller** hydrocarbons (which can include alkenes). It involves breaking the **C–C bonds**.

   You could crack **decane** like this —

$$C_{10}H_{22} \rightarrow C_2H_4 + C_8H_{18}$$
$$\text{decane} \qquad \text{ethene} \quad \text{octane}$$

There are **two types** of **cracking** you need to know about:

THERMAL CRACKING

- It takes place at **high temperature** (up to 1000 °C) and **high pressure** (up to 70 atm).

- It produces a lot of **alkenes**.

- These **alkenes** are used to make heaps of valuable products, like **polymers**. A good example is **poly(ethene)**, which is made from ethene (have a squiz at page 75 for more on polymers).

CATALYTIC CRACKING

- This makes mostly **motor fuels** and **aromatic** hydrocarbons.
  Aromatic compounds contain benzene rings. Benzene rings have six carbon atoms with three double bonds. They're pretty stable because the electrons are delocalised around the carbon ring.

- It uses something called a **zeolite catalyst (hydrated aluminosilicate)**, at a **slight pressure** and **high temperature** (about 450 °C).

- Using a catalyst **cuts costs**, because the reaction can be done at a **lower** temperature and pressure. The catalyst also **speeds** up the reaction, and time is money and all that.

Practice Questions

Q1 What is the general formula for alkanes?

Q2 Draw and name the first four alkanes.

Q3 What's the purpose of the fractional distillation of crude oil?

Q4 What is cracking?

Q5 What type of organic chemical does thermal cracking produce?

Exam Questions

Q1 Crude oil contains many different alkane molecules.
   These are separated using a process called fractional distillation.
   a) Why do the components of crude oil need to be separated?   [1 mark]
   b) What physical property of the molecules is used to separate them?   [1 mark]
   c) A typical alkane found in the petrol (gasoline) fraction has 8 carbon atoms.
      (i) Give the molecular formula for this alkane.   [1 mark]
      (ii) Would you find the petrol fraction near the top or bottom of the fractionating column?
            Explain your answer.   [3 marks]
      (iii) What is the molecular formula of a cycloalkane with 8 carbon atoms.   [1 mark]

Q2 Crude oil is a source of fuels and petrochemicals. It's vaporised and separated into fractions using fractional distillation.
   Some heavier fractions are processed using cracking.
   a) Describe one reason why cracking is carried out.   [2 marks]
   b) Write a possible equation for the thermal cracking of dodecane, $C_{12}H_{26}$.   [1 mark]

Crude oil — not the kind of oil you could take home to meet your mother...

This ain't the most exciting page in the history of the known universe. Although in a galaxy far, far away there may be lots of pages on even more boring topics. But, that's neither here nor there, cos you've got to learn the stuff anyway. Get fractional distillation and cracking straight in your brain and make sure you know why people bother to do it.

Alkanes as Fuels

Alkanes are absolutely fantastic as fuels. Except for the fact that they produce loads of nasty pollutant gases.

Alkanes are Useful Fuels

1) If you burn (**oxidise**) alkanes with **plenty of oxygen**, you get **carbon dioxide** and water — it's a **combustion reaction**.

Here's the equation for the combustion of propane — $C_3H_{8(g)} + 5O_{2(g)} \rightarrow 3CO_{2(g)} + 4H_2O_{(g)}$

This is complete combustion. There's also incomplete combustion, which is really BAD. See below...

2) Alkanes make great fuels — burning just a small amount of **methane** releases a humungous amount **of energy**.

Carbon Monoxide is Formed if Alkanes Burn Incompletely

If there's not enough oxygen, hydrocarbons **combust incompletely**, and you get carbon monoxide gas instead of carbon dioxide. E.g.

$CH_{4(g)} + 1\frac{1}{2}O_{2(g)} \rightarrow CO_{(g)} + 2H_2O_{(g)}$ $C_8H_{18(g)} + 8\frac{1}{2}O_{2(g)} \rightarrow 8CO_{(g)} + 9H_2O_{(g)}$

This is bad news because carbon monoxide gas is poisonous. Carbon monoxide molecules bind to the same sites on **haemoglobin molecules** in red blood cells as oxygen molecules. So **oxygen** can't be carried around the body.

Luckily, carbon monoxide can be removed from exhaust gases by **catalytic converters** on cars.

And if that's Not Bad Enough... Burning Fuels Produces Other Pollutants Too

UNBURNT HYDROCARBONS AND OXIDES OF NITROGEN (NO$_x$) CONTRIBUTE TO SMOG

1) Engines **don't burn** all the fuel molecules. Some of these come out as **unburnt hydrocarbons**.
2) **Oxides of nitrogen** (NO$_x$) are produced when the high pressure and temperature in a car engine cause the nitrogen and oxygen atoms in the air to react together.
3) The hydrocarbons and nitrogen oxides react in the presence of sunlight to form **ground-level ozone** (O_3), which is a major component of **smog**. **ground-level ozone** irritates people's eyes, aggravates respiratory problems and even causes lung damage (ozone isn't nice stuff, unless it is high up in the atmosphere as part of the ozone layer).
4) **Catalytic converters** on cars remove unburnt hydrocarbons and oxides of nitrogen from the exhaust.

SULFUR DIOXIDE

1) **Acid rain** is caused by burning fossil fuels that contain **sulfur**. The sulfur burns to produce **sulfur dioxide** gas which then enters the atmosphere, dissolves in the moisture, and is converted into **sulfuric acid**.
   The same process occurs when nitrogen dioxide escapes into the atmosphere — nitric acid is produced.
2) Acid rain destroys trees and vegetation, as well as corroding buildings and statues and killing fish in lakes. Luckily, sulfur dioxide can be removed from power station flue gases using **calcium oxide**.

Yet More Bad News... Burning Fossil Fuels Contributes to Global Warming

1) The vast majority of scientists believe that **global warming** is caused by increased levels of **carbon dioxide** in the atmosphere due to burning **fossil fuels** (coal, oil and natural gas).
2) Not everyone agrees with this theory, but here's what is true:
   - **Greenhouse gases** stop some of the heat from the Sun from escaping back into space.
     This is the greenhouse effect — it's what keeps the Earth warm enough for us to live here (see the next page).
   - **Carbon dioxide** is a greenhouse gas.
   - Burning **fossil fuels** produces carbon dioxide.
   - The level of carbon dioxide in the atmosphere has **increased** in the last 50 years or so.
   - The average **temperature** of the Earth has **increased dramatically** over the same period.
     This is **global warming**, and it's a big headache for the whole planet.
3) Most scientists have looked at all the **evidence** and agree that the rise in carbon dioxide levels is down to human activity, including burning fossil fuels. They also agree that the extra CO_2 is **enhancing** the greenhouse effect, and that this is the cause of global warming.
4) There are still a few scientists who think that there are **other explanations**, either for the rise in CO_2 levels, or for the cause of global warming. That's part of science — it can take a long time for everyone to accept a theory (and you never know when some new evidence might turn up to prove everyone wrong).

Alkanes as Fuels

Carbon Dioxide isn't the Only Greenhouse Gas

1) Some of the **electromagnetic radiation** from the Sun reaches the Earth and is **absorbed**. The Earth then **re-emits** it as **infrared radiation** (heat).

2) Various gases in the troposphere (the lowest layer of the atmosphere) **absorb** some of this infrared radiation... and **re-emit** it in **all directions** — including back towards Earth, keeping us warm. This is called the '**greenhouse effect**' (even though a real greenhouse doesn't actually work like this, annoyingly).

Visible and UV radiation from the Sun

Some infrared radiation emitted by the Earth is absorbed by greenhouse gases

Some infrared radiation emitted by the Earth escapes

3) The three main greenhouse gases are **water vapour**, **carbon dioxide** and **methane**. **Human activities** have caused a rise in greenhouse gas concentrations, which **enhances** the greenhouse effect. So now **too much heat** is being trapped and the Earth is **getting warmer** — this is **global warming**.

4) Items about global warming on TV and in newspapers usually focus on cutting the levels of **carbon dioxide**, but the other greenhouse gases are important, too.

5) When alkanes in **fossil fuels** are burned they also produce **water vapour**. People tend not to worry so much about water vapour in the atmosphere. There's always been lots of it, and unlike carbon dioxide, the levels have stayed pretty **constant** — and some of it gets removed every time it rains.

6) The other important greenhouse gas is **methane**. Methane's produced by rubbish rotting in **landfill sites**. Methane levels have also risen as we've had to grow more food for our rising population. **Cows** are responsible for large amounts of methane. From both ends.

Vegetarians can't feel entirely smug though. Paddy fields, in which rice is grown, kick out a fair amount of methane too.

Practice Questions

Q1 Which two compounds are produced when an alkane burns completely?

Q2 Why is the incomplete combustion of alkanes a problem?

Q3 Explain how burning fossil fuels may contribute to global warming.

Q4 Name three greenhouse gases.

Exam Questions

Q1 Heptane, C_7H_{16}, is an alkane present in some fuels.

   a) Write a balanced equation for the complete combustion of heptane.   [2 marks]

   b) Fuels often contain oxygenates such as methanol to ensure that the fuel burns completely.

      (i) What toxic compound can be produced by the incomplete combustion of alkanes such as heptane?   [1 mark]

      (ii) Apart from adding oxygenates, how else can this compound be removed from exhaust gases?   [1 mark]

Q2 Burning fossil fuels can cause a variety of environmental problems.

   a) Explain how oxides of nitrogen are produced in car engines.   [2 marks]

   b) Explain why burning fossil fuels in power stations can lead to acid rain and how this problem can be solved.   [3 marks]

Burn, baby, burn — so long as the combustion is complete...

Don't you just hate it when you come up with a great idea, then everyone picks holes in it? Well, just imagine if you were the one who thought of burning alkanes for fuel... it seemed like such a good idea at the time. Despite all the problems, we're still using them — and until we find some suitable alternatives, we all have to deal with the negative consequences.

Enthalpy Changes

A whole new section to enjoy — but don't forget, Big Brother is watching...

Chemical Reactions Usually Have Enthalpy Changes

When chemical reactions happen, there'll be a **change in energy**.
The souped-up chemistry term for this is **enthalpy change** —

> **Enthalpy change**, ΔH (delta H), is the heat energy transferred in a reaction at **constant pressure**. The units of ΔH are **kJ mol^{-1}**.

You write $\Delta H^{\ominus}$ to show that the elements were in their **standard states** and that the measurements were made under **standard conditions**. Standard conditions are **100 kPa (about 1 atm) pressure** and a stated temperature (e.g. ΔH_{298}). In this book, all the enthalpy changes are measured at 298 K (25 °C).

Reactions can be either Exothermic or Endothermic

> **Exothermic** reactions **give out** energy. ΔH is **negative**.

In exothermic reactions, the temperature often goes **up**.

> **Oxidation** is exothermic. Here are two examples:
> - The **combustion** of a fuel like methane ⟹ $CH_{4(g)} + 2O_{2(g)} \rightarrow CO_{2(g)} + 2H_2O_{(l)}$ $\Delta H^{\ominus}_{c,\,298} = -890$ kJ mol^{-1} **exothermic**
>
> - The oxidation of **carbohydrates**, such as glucose, $C_6H_{12}O_6$, in respiration.

> **Endothermic** reactions **absorb** energy. ΔH is **positive**.

In these reactions, the temperature often **falls**.

> The **thermal decomposition** of calcium carbonate is endothermic.
> $$CaCO_{3(s)} \rightarrow CaO_{(s)} + CO_{2(g)} \quad \Delta H^{\ominus}_{r,\,298} = +178 \text{ kJ mol}^{-1} \textbf{ endothermic}$$
>
> The main reactions of **photosynthesis** are also endothermic — sunlight supplies the energy.

Reactions are all about Breaking and Making Bonds

You can only break bonds if you've got enough energy.

When reactions happen, **reactant bonds** are **broken** and **product bonds** are **formed**.
1) You **need** energy to break bonds, so bond breaking is **endothermic (ΔH is positive)**. **Stronger** bonds take **more** energy to break.
2) Energy is **released** when bonds are formed, so this is **exothermic (ΔH is negative)**. **Stronger** bonds release **more** energy when they form.
3) The **enthalpy change** for a reaction is the **overall effect** of these two changes. If you need **more** energy to **break** bonds than is released when bonds are made, ΔH is **positive**. If it's less, ΔH is negative.

Mean Bond Enthalpies are not Exact

Water (H_2O) has got **two O–H bonds**. You'd think it'd take the same amount of energy to break them both... but it **doesn't**.

The **first** bond, H–OH$_{(g)}$: E(H–OH) = +492 kJ mol^{-1}
The **second** bond, H–O$_{(g)}$: E(H–O) = +428 kJ mol^{-1}
(OH$^-$ is a bit easier to break apart because of the extra electron repulsion.)
So, the **mean** bond enthalpy is $\dfrac{492+428}{2}$ = **+460 kJ mol^{-1}**.

The *data book* says the bond enthalpy for O–H is +463 kJ mol^{-1}. It's a bit different because it's the average for a *much bigger range* of molecules, not just water. For example, it includes the O–H bonds in alcohols and carboxylic acids too.

Breaking bonds is always an endothermic process, so mean bond enthalpies are always **positive**.

Enthalpy Changes

Enthalpy Changes Can Be Calculated using Average Bond Enthalpies

In any chemical reaction energy is **absorbed** to **break bonds** and **given out** during **bond formation**.
The difference between the energy absorbed and released is the overall **enthalpy change of reaction**:

Enthalpy Change of Reaction = Total Energy Absorbed – Total Energy Released

Example: Calculate the overall enthalpy change for this reaction:

$N_2 + 3H_2 \rightarrow 2NH_3$

Use the average bond enthalpy values in the table.

Bond	Average Bond Enthalpy
N≡N	945 kJ mol^{-1}
H–H	436 kJ mol^{-1}
N–H	391 kJ mol^{-1}

Bonds broken: 1 × N≡N bond broken = 1 × 945 = 945 kJ mol^{-1}
3 × H–H bonds broken = 3 × 436 = 1308 kJ mol^{-1}

Total Energy Absorbed = 945 + 1308 = **2253 kJ mol^{-1}**

Bonds formed: 6 × N–H bonds formed = 6 × 391 = 2346 kJ mol^{-1}

Total Energy Released = **2346 kJ mol^{-1}**

Now you just subtract 'total energy released' from 'total energy absorbed':

Enthalpy Change of Reaction = 2253 – 2346 = **–93 kJ mol^{-1}**

If you can't remember which value to subtract from which, just take the smaller number from the bigger one then add the sign at the end — positive if 'bonds broken' was the bigger number (endothermic), negative if 'bonds formed' was bigger (exothermic).

There are Different Types of ΔH

1) **Standard enthalpy change of reaction,** $\Delta H_r^\ominus$, is the enthalpy change when the reaction occurs in the **molar quantities** shown in the **chemical equation**, under standard conditions in their standard states.

2) **Standard enthalpy change of formation,** $\Delta H_f^\ominus$, is the enthalpy change when **1 mole** of a **compound** is formed from its **elements** in their standard states under standard conditions, e.g. $2C_{(s)} + 3H_{2(g)} + \frac{1}{2}O_{2(g)} \longrightarrow C_2H_5OH_{(l)}$

3) **Standard enthalpy change of combustion,** $\Delta H_c^\ominus$, is the enthalpy change when **1 mole** of a substance is completely **burned in oxygen** under standard conditions.

Practice Questions

Q1 Explain the terms exothermic and endothermic, giving an example in each case.
Q2 Is energy taken in or released when bonds are broken?
Q3 What is the mean bond enthalpy? How can you work it out?
Q4 Define standard enthalpy of formation and standard enthalpy of combustion.

Exam Questions

Q1 The table shows some average bond enthalpy values.

Bond	C–H	C=O	O=O	O–H
Average Bond Enthalpy (kJ mol^{-1})	435	805	498	464

The complete combustion of methane can be represented by the following equation:

$$CH_{4\,(g)} + 2O_{2\,(g)} \rightarrow CO_{2\,(g)} + 2H_2O_{\,(l)}$$

a) Use the table of bond enthalpies above to calculate the enthalpy change for the reaction. [4 marks]
b) Is the reaction endothermic or exothermic? Explain your answer. [1 mark]

Q2 Methanol, CH_3OH, when blended with petrol, can be used as a fuel. $\Delta H_{c,\,298}^\ominus [CH_3OH] = -726$ kJ mol^{-1}.
a) Write an equation, including state symbols, for the standard enthalpy change of combustion of methanol. [2 marks]
b) Write an equation, including state symbols, for the standard enthalpy change of formation of methanol. [2 marks]
c) Liquid petroleum gas is a fuel that contains propane, C_3H_8.
Explain why the following equation does not represent a standard enthalpy change of combustion. [1 mark]

$$2C_3H_{8(g)} + 10O_{2(g)} \longrightarrow 8H_2O_{(g)} + 6CO_{2(g)} \quad \Delta H_{r,\,298} = -4113 \text{ kJ mol}^{-1}$$

I bonded with my friend — now we're waiting to be surgically separated...

What a lotta definitions. And you need to know them all. If you're going to bother learning them, you might as well do it properly and learn all the pernickety details. They probably seem about as useful as a dead fly in your custard right now, but all will be revealed over the next few pages. Learn them now, so you've got a bit of a head start.

Calculating Enthalpy Changes

Now you know what enthalpy changes are, here's how to calculate them...

You can find out **Enthalpy Changes** in the Lab

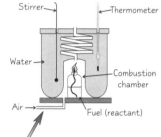

1) To measure the **enthalpy change** for a reaction, you only need to know **two things** —
   • The **number of moles** of the stuff that's reacting.   • The change in **temperature**.

2) How you go about doing the experiment depends on what type of reaction it is.
   Some reactions will quite happily take place in a **container** and you can just stick a
   thermometer in to find out the temperature change. It's best to use a **polystyrene beaker**,
   so that you don't lose or gain much heat through the sides.

3) **Combustion reactions** are trickier because the reactant is burned in air. A **copper calorimeter** containing a **known
   mass of water** is often used. You burn a **known mass of the reactant** and record the **temperature change** of the water.

Calculate **Enthalpy Changes** Using the **Equation $q = mc\Delta T$**

It seems there's a snazzy equation for everything these days, and enthalpy change is no exception —

$q = mc\Delta T$ where, q = heat lost or gained (in joules). This is the same as the enthalpy change if the
pressure is constant.

m = mass of water in the calorimeter, or solution in the polystyrene beaker (in grams)
c = specific heat capacity of water (4.18 J g^{-1}K^{-1})
ΔT = the change in temperature of the water or solution

Example:

In a laboratory experiment, 1.16 g of an organic liquid fuel was completely burned in oxygen.
The heat formed during this combustion raised the temperature of 100 g of water from 295.3 K to 357.8 K.

Calculate the standard enthalpy of combustion, $\Delta H_c^{\ominus}$, of the fuel. Its M_r is 58.

1 First off, you need to calculate the **amount of heat** given out by the fuel using $q = mc\Delta T$.

$q = mc\Delta T$

$q = 100 \times 4.18 \times (357.8 - 295.3) = 26\ 125$ J $= 26.125$ kJ ⟵ *Change the amount of heat from J to kJ.*

Remember — m is the mass of water, NOT the mass of fuel.

2 Next you need to find out **how many moles** of fuel produced this heat. It's back to the old $n = \dfrac{\text{mass}}{M}$ equation.

$$n = \frac{1.16}{58} = 0.02 \text{ moles of fuel}$$

3 The standard enthalpy of combustion involves 1 mole of fuel.

So, the heat produced by 1 mole of fuel $= \dfrac{-26.125}{0.02}$

It's negative because combustion is an exothermic reaction.

$\approx$ **-1306 kJ mol^{-1}**. This is the standard enthalpy change of combustion.

The actual $\Delta H_c^{\ominus}$ of this compound is -1615 kJ mol^{-1} — loads of heat has been **lost** and not measured. E.g. it's likely a
fair bit would escape through the **copper calorimeter** and also the fuel might not **combust completely**.

Hess's Law — *the Total Enthalpy Change is **Independent** of the Route Taken*

Hess's Law says that:

The **total enthalpy change** of a reaction is always **the same**, no matter **which route** is taken.

$2NO_{2(g)} \xrightarrow[\text{Route 1}]{\Delta H_r} N_{2(g)} + 2O_{2(g)}$

+114.4 kJ ⟍   Route 2   ⟋ −180.8 kJ

$2NO_{(g)} + O_{2(g)}$

This law is handy for working out enthalpy changes
that you **can't find directly** by doing an experiment.

Here's an example:
The **total enthalpy change** for route 1 is the **same** as for **route 2**.

So, $\Delta H_r = +114.4 + (-180.8) = -66.4$ kJ mol^{-1}.

Calculating Enthalpy Changes

Enthalpy Changes Can be *Worked Out Indirectly*

The element's being formed from the element, so there's no change

Enthalpy changes of formation are useful for calculating enthalpy changes you can't find directly. You need to know $\Delta H_f^{\ominus}$ for **all** the reactants and products that are **compounds** — the value of $\Delta H_f^{\ominus}$ for elements is **zero**.

Here's how to calculate $\Delta H_r^{\ominus}$ for this reaction: $SO_{2(g)} + 2H_2S_{(g)} \rightarrow 3S_{(s)} + 2H_2O_{(l)}$

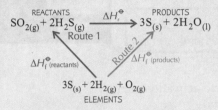

$$\Delta H_f^{\ominus}[SO_{2\,(g)}] = -297 \text{ kJ mol}^{-1}$$

$$\Delta H_f^{\ominus}[H_2S_{(g)}] = -20.2 \text{ kJ mol}^{-1}$$

$$\Delta H_f^{\ominus}[H_2O_{(l)}] = -286 \text{ kJ mol}^{-1}$$

Using **Hess's Law**: Route 1 = Route 2

$\Delta H_r^{\ominus}$ + the sum of $\Delta H_f^{\ominus}$ (reactants) = the sum of $\Delta H_f^{\ominus}$ (products)

So, $\Delta H_r^{\ominus}$ = the sum of $\Delta H_f^{\ominus}$ (**products**) – the sum of $\Delta H_f^{\ominus}$ (**reactants**)

To find $\Delta H_r^{\ominus}$ of this reaction:

Just plug the numbers into the equation above:
$$\Delta H_r^{\ominus} = [0 + (-286 \times 2)] - [-297 + (-20.2 \times 2)] = \textbf{-234.6 kJ mol}^{-1}$$

$\Delta H_f^{\ominus}$ of sulfur is zero — it's an element.

There's 2 moles of H_2O and 2 moles of H_2S.

You can use a similar method to find an enthalpy change from **enthalpy changes of combustion**.

The standard enthalpy changes are all measured at 298 K.

Here's how to calculate $\Delta H_f^{\ominus}$ of **ethanol**...

Using Hess's Law: Route 1 = Route 2

$\Delta H_f^{\ominus}[\text{ethanol}] + \Delta H_c^{\ominus}[\text{ethanol}] = 2\Delta H_c^{\ominus}[C] + 3\Delta H_c^{\ominus}[H_2]$

$\Delta H_f^{\ominus}[\text{ethanol}] + (-1367) = (2 \times -394) + (3 \times -286)$

$\Delta H_f^{\ominus}[\text{ethanol}] = -788 + -858 - (-1367)$

$= \textbf{-279 kJ mol}^{-1}$.

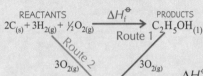

$$\Delta H_c^{\ominus}[C_{(s)}] = -394 \text{ kJ mol}^{-1}$$

$$\Delta H_c^{\ominus}[H_{2\,(g)}] = -286 \text{ kJ mol}^{-1}$$

$$\Delta H_c^{\ominus}[\text{ethanol}_{(l)}] = -1367 \text{ kJ mol}^{-1}$$

Practice Questions

Q1 Briefly describe an experiment that could be carried out to find the enthalpy change of a reaction.

Q2 What equation would you use to calculate the heat change in this experiment?

Q3 Why is the enthalpy change determined in a laboratory likely to be lower than the value shown in a data book?

Q4 What does Hess's Law state?

Q5 What is the standard enthalpy change of formation of any element?

Exam Questions

Q1 Using the facts that (at 298K) the standard enthalpy change of formation of $Al_2O_{3(s)}$ is –1676 kJ mol^{-1} and the standard enthalpy change of formation of $MgO_{(s)}$ is –602 kJ mol^{-1}, calculate the enthalpy change of the following reaction.
$$Al_2O_{3(s)} + 3Mg_{(s)} \rightarrow 2Al_{(s)} + 3MgO_{(s)}$$
[3 marks]

Q2 A 50 cm³ sample of 0.200 M copper(II) sulfate solution placed in a polystyrene beaker gave a temperature increase of 2.6 K when excess zinc powder was added and stirred. Calculate the enthalpy change when 1 mole of zinc reacts. Assume that the specific heat capacity for the solution is 4.18 J g^{-1}K^{-1}. Ignore the increase in volume due to the zinc.

The equation for the reaction is: $\quad Zn_{(s)} + CuSO_{4(aq)} \rightarrow Cu_{(s)} + ZnSO_{4(aq)}$
[8 marks]

To understand this lot, you're gonna need a bar of chocolate. Or two...

To get your head around those Hess diagrams, you're going to have to do more than skim them. You need to be able to use this stuff for any reaction they give you. It'll also help if you know the definitions for those standard enthalpy thingumabobs on page 45. If you didn't bother learning them, have a quick flick back and remind yourself about them.

Reaction Rates and Catalysts

The rate of a reaction is just how quickly it happens. Lots of things can make it go faster or slower.

Particles **Must** Collide to **React**

1)   Particles in liquids and gases are **always moving** and **colliding** with **each other**.
They **don't** react every time though — only when the **conditions** are right.
A reaction **won't** take place between two particles **unless** —

>   • They collide in the **right direction**. They need to be **facing** each other the right way.
>   • They collide with at least a certain **minimum** amount of kinetic (movement) **energy**.

This stuff's called **Collision Theory**.

2)   The **minimum amount of kinetic energy** particles need to react is called the **activation energy**.
The particles need this much energy to **break the bonds** to start the reaction.

3)   Reactions with **low activation energies** often happen **pretty easily**. But reactions with
high activation energies don't. You need to give the particles extra energy by **heating** them.

To make this a bit clearer, here's an **enthalpy profile diagram**.

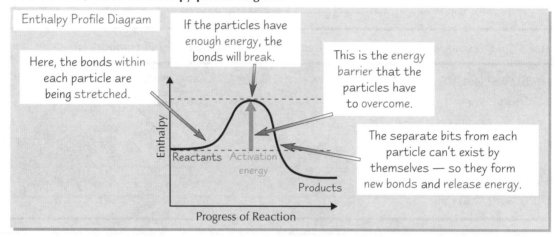

Enthalpy Profile Diagram

Here, the bonds within each particle are being stretched.

If the particles have enough energy, the bonds will break.

This is the energy barrier that the particles have to overcome.

The separate bits from each particle can't exist by themselves — so they form new bonds and release energy.

Enthalpy

Reactants   Activation energy

Products

Progress of Reaction

Molecules in a Gas **Don't** all have the **Same Amount of Energy**

Imagine looking down on Oxford Street when it's teeming with people. You'll see some people
ambling along **slowly**, some hurrying **quickly**, but most of them will be walking with a **moderate speed**.
It's the same with the **molecules** in a gas. Some **don't have much kinetic energy** and move **slowly**.
Others have **loads of kinetic energy** and **whizz** along. But most molecules are somewhere **in between**.

If you plot a **graph** of the **numbers of molecules** in a gas with different **kinetic energies** you get a
Maxwell-Boltzmann distribution. It looks like this —

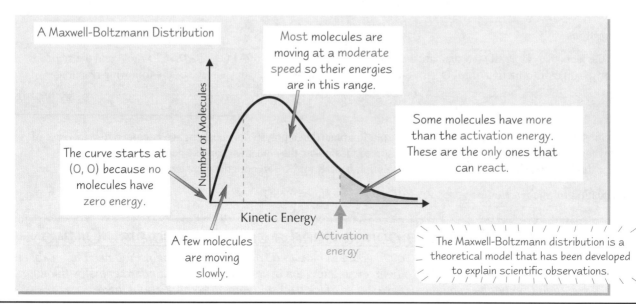

A Maxwell-Boltzmann Distribution

Most molecules are moving at a moderate speed so their energies are in this range.

Some molecules have more than the activation energy. These are the only ones that can react.

The curve starts at (0, 0) because no molecules have zero energy.

Number of Molecules

Kinetic Energy

A few molecules are moving slowly.

Activation energy

The Maxwell-Boltzmann distribution is a theoretical model that has been developed to explain scientific observations.

Reaction Rates and Catalysts

Increasing the Temperature makes Reactions Faster

1) If you increase the **temperature**, the particles will on average have more **kinetic energy** and will move **faster**.
2) So, a **greater proportion** of molecules will have at least the **activation energy** and be able to **react**. This changes the **shape** of the **Maxwell-Boltzmann distribution curve** — it pushes it over to the **right**.

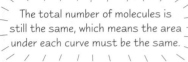

The total number of molecules is still the same, which means the area under each curve must be the same.

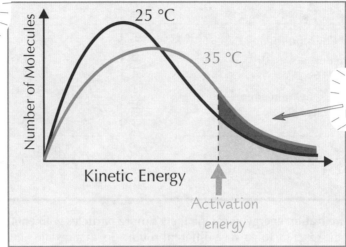

At higher temperatures, more molecules have at least the activation energy.

3) Because the molecules are flying about **faster**, they'll **collide more often**. This is **another reason** why increasing the temperature makes a reaction faster. So,

> **Small temperature increases** can lead to **large increases in reaction rate**.

Increasing Concentration also Increases the Rate of Reaction

1) If you increase the **concentration** of reactants in a **solution**, the particles will on average be **closer together**.
2) If they're closer, they'll **collide more often**. If there are **more collisions**, they'll have **more chances** to react.
3) If the reaction involves gases, increasing the **pressure** of the gases works in just the same way.

Catalysts Increase the Rate of Reactions Too

You can use **catalysts** to make chemical reactions happen **faster**. Learn this definition:

> A **catalyst** increases the **rate** of a reaction by providing an **alternative reaction pathway** with a **lower activation energy**. The catalyst is **chemically unchanged** at the end of the reaction.

There's more on this on the next page. Bet you can't wait.

1) Catalysts are **great**. They **don't** get used up in reactions, so you only need a **tiny bit** of catalyst to catalyse a **huge** amount of stuff. They **do** take part in reactions, but they're **remade** at the end.
2) Catalysts are **very fussy** about which reactions they catalyse. Many will usually **only** work on a single reaction.
3) Catalysts **save heaps of money** in industrial processes.

Reaction Rates and Catalysts

Enthalpy Profiles and Boltzmann Distributions Show Why Catalysts Work

If you look at an **enthalpy profile** together with a **Maxwell-Boltzmann Distribution**, you can see **why** catalysts work.

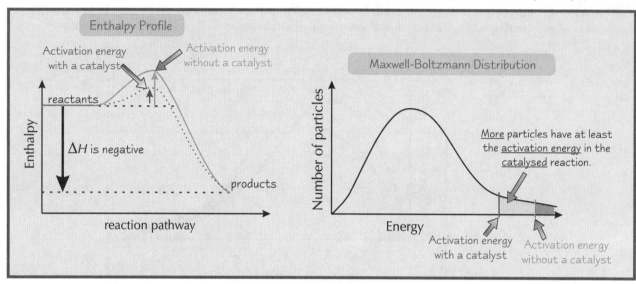

The catalyst **lowers the activation energy**, meaning there's **more particles** with **enough energy** to react when they collide. It does this by allowing the reaction to go **via a different route**. So, in a certain amount of time, **more particles react**.

Practice Questions

Q1 Does every particle collision result in a reaction? Explain your answer?

Q2 Explain the term 'activation energy'.

Q3 What does a Maxwell-Boltzmann distribution show?

Q4 Sketch a Maxwell-Boltzmann distribution for molecules at two different temperatures.

Q5 Explain what a catalyst is.

Q6 Draw a Maxwell-Boltzmann distribution diagram to show how a catalyst works.

Exam Questions

Q1 Nitrogen oxide (NO) and ozone (O_3) react to produce nitrogen dioxide (NO_2) and oxygen (O_2).
The collision between the two molecules does not always lead to a reaction, even if the molecules are
orientated correctly. Explain why this is. [1 mark]

Q2 Use collision theory to explain why the reaction between
a solid and a liquid is generally faster than that between two solids. [2 marks]

Q3 The decomposition of hydrogen peroxide, H_2O_2, into water and oxygen is catalysed by manganese(IV) oxide, MnO_2.
a) Write a balanced equation for the reaction. [2 marks]
b) Sketch a Maxwell-Boltzmann distribution for the reaction.
Mark on the activation energy for the catalysed and uncatalysed process. [3 marks]
c) Referring to your diagram from part b), explain how manganese(IV) oxide acts as a catalyst. [3 marks]
d) What would be the effect of raising the temperature of this reaction? Explain the effect. [3 marks]

I'm a catalyst — I like to speed up arguments without getting too involved...

Whatever you do, don't confuse the effect of catalysts with the effect of a temperature change. They both mean more particles have the activation energy. Catalysts do this by lowering the activation energy, BUT increasing temperature does this by giving the particles more energy. Get these mixed up and you'll be the laughing stock of the Examiners' tea room.

Reversible Reactions

There's a lot of to-ing and fro-ing on this page. Mind your head doesn't start spinning.

Reversible Reactions Can Reach Dynamic Equilibrium

1) Lots of chemical reactions are **reversible** — they go **both ways**. To show a reaction's reversible, you stick in a $\rightleftharpoons$. Here's an example:

$$H_{2(g)} + I_{2(g)} \rightleftharpoons 2HI_{(g)}$$

This reaction can go in **either direction** —

forwards $H_{2(g)} + I_{2(g)} \rightarrow 2HI_{(g)}$ ....or backwards $2HI_{(g)} \rightarrow H_{2(g)} + I_{2(g)}$.

2) As the **reactants** get used up, the **forward** reaction **slows down** — and as more **product** is formed, the **reverse** reaction **speeds up**.

3) After a while, the forward reaction will be going at exactly the **same rate** as the backward reaction. The amounts of reactants and products **won't be changing** any more, so it'll seem like **nothing's happening**. It's a bit like you're **digging a hole**, while someone else is **filling it in** at exactly the **same speed**. This is called a **dynamic equilibrium**.

4) A **dynamic equilibrium** can only happen in a **closed system**. This just means nothing can get in or out.

Le Chatelier's Principle Predicts what will Happen if Conditions are Changed

If you **change** the **concentration**, **pressure** or **temperature** of a reversible reaction, you're going to **alter** the **position of equilibrium**. This just means you'll end up with **different amounts** of reactants and products at equilibrium.

If the position of equilibrium moves to the **left**, you'll get more **reactants**.

$$H_{2(g)} + I_{2(g)} \rightleftharpoons 2HI_{(g)}$$

If the position of equilibrium moves to the **right**, you'll get more **products**.

$$H_{2(g)} + I_{2(g)} \rightleftharpoons 2HI_{(g)}$$

Mr and Mrs Le Chatelier celebrate another successful year in the principle business

Le Chatelier's principle tells you how the **position of equilibrium** will change if a **condition changes**:

If there's a change in **concentration**, **pressure** or **temperature**, the equilibrium will move to help **counteract** the change.

So, basically, if you **raise the temperature**, the position of equilibrium will shift to try to **cool things down**. And, if you **raise the pressure or concentration**, the position of equilibrium will shift to try to **reduce it again**.

Catalysts Don't Affect The Position of Equilibrium

Catalysts have **NO EFFECT** on the **position of equilibrium**.
They **can't** increase **yield** — but they **do** mean equilibrium is reached **faster**.

Reversible Reactions

Here's Some Handy Rules for Using Le Chatelier's Principle

CONCENTRATION $2SO_{2(g)} + O_{2(g)} \rightleftharpoons 2SO_{3(g)}$

1) If you **increase** the **concentration** of a **reactant** (SO_2 or O_2), the equilibrium tries to **get rid** of the extra reactant. It does this by making **more product** (SO_3). So the equilibrium's shifted to the **right**.

2) If you **increase** the **concentration** of the **product** (SO_3), the equilibrium tries to remove the extra product. This makes the **reverse reaction** go faster. So the equilibrium shifts to the **left**.

3) **Decreasing** the concentrations has the **opposite effect**.

PRESSURE   (changing this only affects **equilibria involving gases**)

1) **Increasing** the pressure shifts the equilibrium to the side with **fewer** gas molecules. This **reduces** the pressure.

2) **Decreasing** the pressure shifts the equilibrium to the side with **more** gas molecules. This **raises** the pressure again.

> There's 3 moles on the left, but only 2 on the right. $\Longrightarrow$ $2SO_{2(g)} + O_{2(g)} \rightleftharpoons 2SO_{3(g)}$
> So, an increase in pressure shifts the equilibrium to the right.

TEMPERATURE

1) **Increasing** the temperature means **adding heat**.
   The equilibrium shifts in the **endothermic (positive ΔH) direction** to absorb this heat.

2) **Decreasing** the temperature **removes heat**.
   The equilibrium shifts in the **exothermic (negative ΔH) direction** to try to replace the heat.

3) If the forward reaction's **endothermic**, the reverse reaction will be **exothermic**, and vice versa.

> This reaction's exothermic in the forward direction.   Exothermic $\Longrightarrow$
> If you increase the temperature, the equilibrium $2SO_{2(g)} + O_{2(g)} \rightleftharpoons 2SO_{3(g)}$ $\Delta H = -197$ kJ mol^{-1}
> shifts to the left to absorb the extra heat. $\Longleftarrow$ Endothermic

Right, you've got to be able to **apply** this Le Chatelier's Principle stuff to industrial processes —
like the production of **ethanol** and **methanol**.

Ethanol can be formed from Ethene and Steam

1) **Ethanol** is produced via a **reversible exothermic reaction** between **ethene** and **steam**:

$$C_2H_{4(g)} + H_2O_{(g)} \rightleftharpoons C_2H_5OH_{(g)} \qquad \Delta H = -46 \text{ kJ mol}^{-1}$$

2) The reaction is carried out at a pressure of **60-70 atmospheres** and a temperature of **300 °C**, with a catalyst of **phosphoric acid**.

The Conditions Chosen are a Compromise

1) Because it's an **exothermic reaction**, **lower** temperatures favour the forward reaction.
   This means that at lower temperatures **more** ethane and steam is converted to ethanol — you get a better **yield**.

2) But **lower temperatures** mean a **slower rate of reaction**. You'd be **daft** to try to get a **really high yield** of ethanol if it's going to take you 10 years. So the 300 °C is a **compromise** between **maximum yield** and **a faster reaction**.

3) **Higher pressures** favour the **forward reaction**, so a pressure of **60-70 atmospheres** is used — **high pressure** moves the reaction to the side with **fewer molecules of gas**. **Increasing the pressure** also increases the **rate** of reaction.

4) Cranking up the pressure as high as you can sounds like a great idea so far. But **high pressures** are **expensive** to produce. You need **stronger pipes** and **containers** to withstand high pressure. And, in this process, increasing the pressure can also cause **side reactions** to occur.

5) So the **60-70 atmospheres** is a **compromise** between **maximum yield** and **expense**.
   In the end, it all comes down to **minimising costs**.

Recycling Unreacted Ethene Also Saves Money

1) Only a **small proportion** of the ethene reacts each time the gases pass through the catalyst.

2) To save money and raw materials, the **unreacted ethene** is separated from the liquid ethanol and **recycled** back into the reactor. Thanks to this around **95%** of the ethene is eventually converted to ethanol.

Reversible Reactions

Methanol can be Produced from Hydrogen and Carbon Monoxide

1) **Methanol** is also made industrially in a **reversible reaction**. It's made from **hydrogen** and **carbon monoxide**:

$$2H_{2(g)} + CO_{(g)} \rightleftharpoons CH_3OH_{(g)} \qquad \Delta H = -90 \text{ kJ mol}^{-1}$$

Industrial conditions — **pressure**: 50-100 atmospheres, **temperature**: 250 °C, **catalyst**: mixture of copper, zinc oxide and aluminium oxide

2) Just like with the production of **ethanol**, the conditions used are a **compromise** between keeping **costs** low and **yield** high.

Methanol and Ethanol are Important Liquid Fuels

Methanol and ethanol are used as fuels in some forms of motor racing.

1) Methanol is mainly used to make other chemicals, but both **methanol** and **ethanol** can also be used as **fuels for cars** — either on their own, or added to petrol.

2) Ethanol and methanol are thought of as **greener** than petrol — they can be made from **renewable resources** and they produce **fewer pollutants** (like NO_x and CO).

3) Methanol and ethanol can both be **carbon neutral fuels** (pretty much). See page 76 for more on why ethanol is thought of as carbon neutral.

Something is _carbon neutral_ if it has no net annual carbon (greenhouse gas) emissions to the atmosphere.

Practice Questions

Q1 Using an example, explain the terms 'reversible' and 'dynamic equilibrium'.

Q2 If the equilibrium moves to the right, do you get more products or reactants?

Q3 A reaction at equilibrium is endothermic in the forward direction. What happens to the position of equilibrium as the temperature is increased?

Q4 Write down the equation for making ethanol from ethene and steam.

Q5 What does 'carbon neutral' mean?

Exam Questions

Q1 Nitrogen and oxygen gases were reacted together in a closed flask and allowed to reach equilibrium with the nitrogen monoxide formed. The forward reaction is endothermic.

$$N_{2(g)} + O_{2(g)} \rightleftharpoons 2NO_{(g)}$$

    a) State Le Chatelier's principle. [1 mark]

    b) Explain how the following changes would affect the position of equilibrium of the above reaction:
       (i)   Pressure is **increased**. [2 marks]
       (ii) Temperature is **reduced**. [2 marks]
       (iii) Nitrogen monoxide is removed. [1 mark]

    c) What would be the effect of a catalyst on the composition of the equilibrium mixture? [1 mark]

Q2 The manufacture of ethanol can be represented by the reaction: $C_2H_{4(g)} + H_2O_{(g)} \rightleftharpoons C_2H_5OH_{(g)}$ $\Delta H = -46 \text{ kJ mol}^{-1}$
Typical conditions are 300 °C and 60-70 atmospheres.

    a) Explain, in molecular terms, why a temperature lower than the one quoted is not used. [3 marks]

    b) Explain why a pressure higher than the one quoted is not often used. [2 marks]

Only going forward cos we can't find reverse...

Equilibria never do what you want them to do. They always **oppose** you. Be sure you know what happens to an equilibrium if you change the conditions. A word about pressure — if there's the same number of gas moles on each side of the equation, then you can raise the pressure as high as you like and it won't make a blind bit of difference to the position of equilibrium.

Redox Reactions

This double page has more occurences of "oxidation" than the Beatles' "All You Need is Love" features the word "love".

If Electrons are Transferred, it's a Redox Reaction

1) A **loss** of electrons is called **oxidation**. A **gain** in electrons is called **reduction**.
2) Reduction and oxidation happen **simultaneously** — hence the term "**redox**" reaction.
3) An **oxidising agent accepts** electrons and gets reduced.
4) A **reducing agent donates** electrons and gets oxidised.

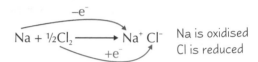

$$Na + \tfrac{1}{2}Cl_2 \xrightarrow{\quad -e^- \quad} Na^+ Cl^- \xleftarrow{\quad +e^- \quad}$$

Na is oxidised
Cl is reduced

Sometimes it's easier to talk about Oxidation States

(It's also called oxidation <u>number</u>.)

There are lots of rules. Take a deep breath...

1) All atoms are treated as **ions** for this, even if they're covalently bonded.

2) Uncombined **elements** have an oxidation state of **0**.

3) Elements just bonded to **identical atoms**, like O_2 and H_2, also have an oxidation state of **0**.

4) The oxidation state of a simple **monatomic ion**, e.g. Na^+, is the same as its **charge**.

5) In **compounds** or **compound ions**, the **overall oxidation state** is just the ion charge.

SO_4^{2-} — **overall oxidation state = –2**,
oxidation state of **O = –2** (total = –8),
so oxidation state of **S = +6**

> Within an ion, the most electronegative element has a negative oxidation state (equal to its ionic charge). Other elements have more positive oxidation states.

6) The sum of the oxidation states for a **neutral compound** is **0**.

Fe_2O_3 — **overall oxidation state = 0**, oxidation state of **O = –2**
(total = –6), so oxidation state of **Fe = +3**

7) Combined **oxygen** is nearly always –2, except in peroxides, where it's –1,
(and in the fluorides OF_2, where it's +2, and O_2F_2, where it's +1 (and O_2 where it's 0).

In H_2O, oxidation state of **O = –2**, but in H_2O_2, oxidation state of **H** has to be **+1** (an H atom can only lose one electron), so oxidation state of **O = –1**

8) Combined **hydrogen** is +1, except in metal hydrides where it is –1 (and H_2 where it's 0).

In **HF**, oxidation state of **H = +1**, but in **NaH**, oxidation state of **H = –1**

Roman Numerals Give Oxidation States

Sometimes, oxidation states aren't clear from the formula of a compound.

If you see **Roman numerals** in a chemical name, it's an **oxidation number**.

E.g. copper has oxidation state **+2** in **copper(II) sulfate** and
manganese has oxidation state **+7** in a **manganate(VII) ion** (MnO_4^-)

Hands up if you like
Roman numerals...

Redox Reactions

You can Write Half-Equations and Combine them into Redox Equations

1) **Ionic half-equations** show oxidation or reduction.
2) You can **combine** half-equations for different oxidising or reducing agents together to make **full equations** for redox reactions.

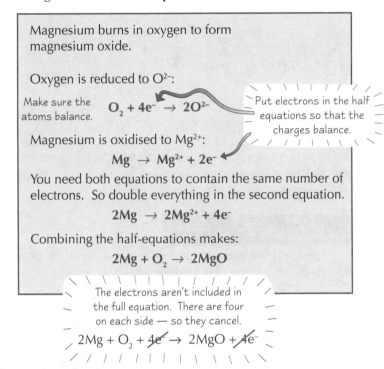

Magnesium burns in oxygen to form magnesium oxide.

Oxygen is reduced to O^{2-}:

Make sure the atoms balance. $O_2 + 4e^- \rightarrow 2O^{2-}$

Put electrons in the half equations so that the charges balance.

Magnesium is oxidised to Mg^{2+}:

$$Mg \rightarrow Mg^{2+} + 2e^-$$

You need both equations to contain the same number of electrons. So double everything in the second equation.

$$2Mg \rightarrow 2Mg^{2+} + 4e^-$$

Combining the half-equations makes:

$$2Mg + O_2 \rightarrow 2MgO$$

The electrons aren't included in the full equation. There are four on each side — so they cancel.

$$2Mg + O_2 + 4e^- \rightarrow 2MgO + 4e^-$$

Aluminum reacts with chlorine to form aluminium chloride.

Aluminium is oxidised to Al^{3+}:

$$Al \rightarrow Al^{3+} + 3e^-$$

Make sure the atoms and charges balance.

Chlorine is reduced to Cl^-:

$$Cl_2 + 2e^- \rightarrow 2Cl^-$$

Now make sure the equations each contain the same number of electrons.

$$Al \rightarrow Al^{3+} + 3e^- \xrightarrow{x2} 2Al \rightarrow 2Al^{3+} + 6e^-$$

$$Cl_2 + 2e^- \rightarrow 2Cl^- \xrightarrow{x3} 3Cl_2 + 6e^- \rightarrow 6Cl^-$$

Combining the half-equations makes:

$$2Al + 3Cl_2 \rightarrow 2AlCl_3$$

Practice Questions

Q1 What is a reducing agent?

Q2 What is the usual oxidation state of oxygen combined with another element?

Q3 What is the oxidation state of hydrogen in H_2 gas?

Q4 What always appears in a half-equation that does not appear in a full equation?

Exam Questions

Q1 Lithium oxide forms when lithium is burned in air.
The equation for the combustion of lithium is: $4Li_{(s)} + O_{2\,(g)} \rightarrow 2Li_2O_{(s)}$

  a) Define oxidation and reduction in terms of the movement of electrons. [1 mark]

  b) What is the oxidation state of lithium in:
     (i) Li   (ii) Li_2O [2 marks]

  c) Write half-equations for the reaction of lithium with oxygen.
     State which reactant is oxidised and which is reduced. [5 marks]

Q2 Halogens are powerful oxidising agents.
The half-equation for chlorine acting as an oxidising agent is: $Cl_2 + 2e^- \rightarrow 2Cl^-$

  a) Define the term oxidising agent in terms of electron movement. [1 mark]

  b) Write a balanced half-equation for the oxidation of indium metal to form In^{3+} ions. [2 marks]

  c) Use your answer to b) and the equation above to form a balanced equation
     for the reaction of indium with chlorine by combining half-equations. [2 marks]

Redox — relax in a lovely warm bubble bath...

Ionic equations are so evil even Satan wouldn't mess with them. But they're on the syllabus, so you can't ignore them. Have a flick back to p16 if they're freaking you out.

And while we're on the oxidation page, I suppose you ought to learn the most famous memory aid thingy in the world...

OIL RIG
- **Oxidation Is Loss**
- **Reduction Is Gain**
(of electrons)

Group 7 — The Halogens

Here comes a page jam-packed with golden nuggets of halogen fun. Oh yes, I kid you not.
This page is the Alton Towers of AS Chemistry... white-knuckle excitement all the way...

The word halogen should be used when describing the atom (X) or molecule (X_2), but the word halide is used to describe the negative ion (X^-).

Halogens are the Highly Reactive Non-Metals of Group 7

The table below gives some of the main properties of the first 4 halogens.

halogen	formula	colour	physical state	electronic structure	electronegativity
fluorine	F_2	pale yellow	gas	$1s^2\ 2s^2\ 2p^5$	increases
chlorine	Cl_2	green	gas	$1s^2\ 2s^2\ 2p^6\ 3s^2\ 3p^5$	up
bromine	Br_2	red-brown	liquid	$1s^2\ 2s^2\ 2p^6\ 3s^2\ 3p^6\ 3d^{10}\ 4s^2\ 4p^5$	the
iodine	I_2	grey	solid	$1s^2\ 2s^2\ 2p^6\ 3s^2\ 3p^6\ 3d^{10}\ 4s^2\ 4p^6\ 4d^{10}\ 5s^2\ 5p^5$	group

1) **Their boiling points increase down the group**
    This is due to the increasing strength of the **Van der Waals forces** as the size and relative mass of the atoms increases. This trend is shown in the changes of **physical state** from fluorine (gas) to iodine (solid).

2) **Electronegativity decreases down the group**.
    Electronegativity, remember, is the tendency of an atom to **attract** a bonding pair of **electrons**. The halogens are all highly electronegative elements. But larger atoms attract electrons **less** than smaller ones. So, going down the group, as the atoms become **larger**, the electronegativity **decreases**.

Halogens Displace Less Reactive Halide Ions from Solution

1) When the halogens react, they **gain an electron**. They get **less reactive down the group**, because the atoms become larger (and less electronegative). So you can say that the halogens become **less oxidising** down the group.

2) The **relative oxidising strengths** of the halogens can be seen in their **displacement reactions** with the halide ions: ⟹

	Potassium chloride solution $KCl_{(aq)}$ - colourless	Potassium bromide solution $KBr_{(aq)}$ - colourless	Potassium iodide solution $KI_{(aq)}$ - colourless
Chlorine water $Cl_{2(aq)}$ - colourless	no reaction	orange solution (Br_2) formed	brown solution (I_2) formed
Bromine water $Br_{2(aq)}$ - orange	no reaction	no reaction	brown solution (I_2) formed
Iodine solution $I_{2(aq)}$ - brown	no reaction	no reaction	no reaction

A **halogen** will **displace a halide** from solution if the halide is **below it** in the periodic table.

3) These displacement reactions can be used to help **identify** which halogen (or halide) is present in a solution.

Halogen	Displacement reaction	Ionic equation
Cl	chlorine (Cl_2) will displace bromide (Br^-) and iodide (I^-)	$Cl_{2(aq)} + 2Br^-_{(aq)} \rightarrow 2Cl^-_{(aq)} + Br_{2(aq)}$ $Cl_{2(aq)} + 2I^-_{(aq)} \rightarrow 2Cl^-_{(aq)} + I_{2(aq)}$
Br	bromine (Br_2) will displace iodide (I^-)	$Br_{2(aq)} + 2I^-_{(aq)} \rightarrow 2Br^-_{(aq)} + I_{2(aq)}$
I	no reaction with F^-, Cl^-, Br^-	

Chlorine and Sodium Hydroxide make Bleach

If you mix chlorine gas with dilute sodium hydroxide at **room temperature**, you get **sodium chlorate(I) solution**, $NaClO_{(aq)}$, which just happens to be common household **bleach**.

Ox. state:
$$2NaOH_{(aq)} + Cl_{2(aq)} \rightarrow NaClO_{(aq)} + NaCl_{(aq)} + H_2O_{(l)}$$
$$0 \qquad\qquad +1 \qquad -1$$

The oxidation state of Cl goes up **and** down. This is **disproportionation**.

ClO^- is the chlorate(I) ion.
Chloride's oxidation state is +1 in this ion.

The sodium chlorate(I) solution (bleach) has loads of uses — it's used in **water treatment**, to bleach **paper** and **textiles**... and it's good for **cleaning toilets**, too. Handy...

Group 7 — The Halogens

Chlorine is used to Kill Bacteria in Water

When you mix chlorine with water, it undergoes disproportionation.

You end up with a mixture of hydrochloric acid and **chloric(I) acid** (also called hypochlorous acid).

Aqueous chloric(I) acid **ionises** to make **chlorate(I) ions** (also called hypochlorite ions).

Chlorate(I) ions **kill bacteria**.

$$Cl_{2(g)} + H_2O_{(l)} \rightleftharpoons HCl_{(aq)} + HClO_{(aq)}$$

Ox. No. of Cl: $\quad 0 \qquad\qquad\qquad -1 \qquad\qquad +1$

$\qquad\qquad\qquad\qquad\qquad\qquad$ hydrochloric acid $\quad$ chloric(I) acid

$$HClO_{(aq)} + H_2O_{(l)} \rightleftharpoons ClO^-_{(aq)} + H_3O^+_{(aq)}$$

So, **adding chlorine** (or a compound containing chlorate(I) ions) to water can make it safe to **drink** or **swim** in.

1) In the UK our drinking water is **treated** to make it safe. **Chlorine** is an important part of water treatment:

- It **kills disease-causing microorganisms**.
- Some chlorine persists in the water and **prevents reinfection** further down the supply.
- It prevents the growth of **algae**, eliminating **bad tastes** and **smells**, and **removes discolouration** caused by organic compounds.

2) However, there are risks from using chlorine to treat water:

- **Chlorine gas** is **very harmful** if it's breathed in — it irritates the **respiratory system**. **Liquid chlorine** on the skin or eyes causes severe **chemical burns**. Accidents involving chlorine could be really serious, or fatal.
- Water contains a variety of organic compounds, e.g. from the decomposition of plants. Chlorine reacts with these compounds to form **chlorinated hydrocarbons**, e.g. chloromethane (CH_3Cl) — and many of these chlorinated hydrocarbons are carcinogenic (cancer-causing). However, this increased cancer risk is small compared to the risks from untreated water — a cholera epidemic, say, could kill thousands of people.

3) There are ethical considerations too. We don't get a **choice** about having our water chlorinated — some people object to this as forced 'mass medication'.

Practice Questions

Q1 Place the halogens F, Cl, Br and I in order of increasing: (a) boiling point, (b) electronegativity.

Q2 What would be seen when chlorine water is added to potassium iodide solution?

Q3 How is common household bleach formed?

Q4 Write the equation for the reaction of chlorine with water. State underneath the oxidation numbers of chlorine.

Exam Questions

Q1 a) Write an ionic equation for the reaction between iodine solution and sodium astatide (NaAt). [1 mark]

    b) For the equation in (a), deduce which substance is oxidised. [1 mark]

Q2 a) Describe and explain the trends in:
      (i) the boiling points of Group 7 elements as you move down the Periodic Table. [3 marks]
      (ii) electronegativity of Group 7 elements as you move down the Periodic Table. [3 marks]

    b) Which halogen is the most powerful oxidising agent? [1 mark]

Q3 Chlorine is added to the water in public swimming baths in carefully controlled quantities.

    a) Write an equation for the reaction of chlorine with water [2 marks]

    b) Why is chlorine added to the water in swimming baths, and why must the quantity added be carefully controlled? [2 marks]

Don't skip this page — it could cost you £31 000...

Let me explain... the other night I was watching Who Wants to Be a Millionaire, and this question was on for £32 000:

> Which of the these elements is a halogen?
> A Argon   B Nitrogen
> C Fluorine   D Sodium

Bet Mr Redmond from Wiltshire wishes he paid more attention in Chemistry now, eh. Ha sucker...

Halide Ions

OK, a quick reminder of the basics first. Halides are compounds with the −1 halogen ion (e.g. Cl⁻, Br⁻, I⁻) like KI, HCl, NaBr. They all end in "-ide" — chloride, bromide, iodide. Got that? Good. Now, you're ready to go in...

The **Reducing Power** of Halides **Increases** Down the Group...

To reduce something, the halide ion needs to lose an electron from its outer shell.
How easy this is depends on the **attraction** between the **nucleus** and the outer **electrons**.

As you go down the group, the attraction gets **weaker** because:

> 1) the ions get bigger, so the electrons are **further** away from the positive nucleus
> 2) there are extra inner electron shells, so there's a greater **shielding** effect.

An example of them doing this is the good old halogen / halide displacement reaction
(the one you learned on p56... yes, that one). And here comes some more examples to learn...

...which Explains their Reactions with **Sulfuric Acid**

All the halides react with concentrated sulfuric acid to give a **hydrogen halide** as a
product to start with. But what happens next depends on which halide you've got...

Reaction of NaF or NaCl with H_2SO_4

$$NaF_{(s)} + H_2SO_{4(aq)} \rightarrow NaHSO_{4(s)} + HF_{(g)}$$

$$NaCl_{(s)} + H_2SO_{4(aq)} \rightarrow NaHSO_{4(s)} + HCl_{(g)}$$

1) Hydrogen fluoride (HF) or hydrogen chloride gas (HCl) is formed. You'll see misty fumes as the gas comes into contact with moisture in the air.

2) But HF and HCl aren't strong enough reducing agents to reduce the sulfuric acid, so the reaction stops there.

3) It's not a redox reaction — the oxidation states of the halide and sulfur stay the same (−1 and +6).

Reaction of NaBr with H_2SO_4

$$NaBr_{(s)} + H_2SO_{4(aq)} \rightarrow NaHSO_{4(s)} + HBr_{(g)}$$

$$2HBr_{(aq)} + H_2SO_{4(aq)} \rightarrow Br_{2(g)} + SO_{2(g)} + 2H_2O_{(l)}$$

ox. state of S:		+6	→ +4	reduction
ox. state of Br:	−1		→ 0	oxidation

1) The first reaction gives misty fumes of hydrogen bromide gas (HBr).

2) But the HBr is a stronger reducing agent than HCl and reacts with the H_2SO_4 in a redox reaction.

3) The reaction produces choking fumes of SO_2 and orange fumes of Br_2.

Reaction of NaI with H_2SO_4

$$NaI_{(s)} + H_2SO_{4(aq)} \rightarrow NaHSO_{4(s)} + HI_{(g)}$$

$$2HI_{(g)} + H_2SO_{4(aq)} \rightarrow I_{2(s)} + SO_{2(g)} + 2H_2O_{(l)}$$

ox. state of S:		+6	→ +4	reduction
ox. state of I:	−1		→ 0	oxidation

1) Same initial reaction giving HI gas.

2) The HI then reduces H_2SO_4 like above.

3) But HI (being well 'ard as far as reducing agents go) keeps going and reduces the SO_2 to H_2S.

$$6HI_{(g)} + SO_{2(g)} \rightarrow H_2S_{(g)} + 3I_{2(s)} + 2H_2O_{(l)}$$

ox. state of S:		+4	→ −2	reduction
ox. state of I:	−1		→ 0	oxidation

> H_2S gas is toxic and smells of bad eggs.
> A bit like my mate Andy at times...

Halide Ions

Silver Nitrate Solution is used to Test for Halides

The test for halides is dead easy. First you add **dilute nitric acid** to remove ions which might interfere with the test. Then you just add **silver nitrate solution** ($AgNO_{3\ (aq)}$). A **precipitate** is formed (of the silver halide).

$$Ag^+_{(aq)} + X^-_{(aq)} \rightarrow AgX_{(s)} \qquad \text{...where X is F, Cl, Br or I}$$

1) The **colour** of the precipitate identifies the halide.

SILVER NITRATE TEST FOR HALIDE IONS...

Fluoride F⁻:	no precipitate
Chloride Cl⁻:	white precipitate
Bromide Br⁻:	cream precipitate
Iodide I⁻:	yellow precipitate

2) Then to be extra sure, you can test your results by adding **ammonia solution**. Each silver halide has a different solubility in ammonia.

SOLUBILITY OF SILVER HALIDE PRECIPITATES IN AMMONIA...

Chloride Cl⁻:	white precipitate, dissolves in dilute $NH_{3(aq)}$
Bromide Br⁻:	cream precipitate, dissolves in conc. $NH_{3(aq)}$
Iodide I⁻:	yellow precipitate, insoluble in conc. $NH_{3(aq)}$

Practice Questions

Q1 Give two reasons why a bromide ion is a more powerful reducing agent than a chloride ion.

Q2 Name the gaseous products formed when sodium bromide reacts with concentrated sulfuric acid.

Q3 What is produced when potassium iodide reacts with concentrated sulfuric acid?

Q4 How would you test whether an aqueous solution contained chloride ions?

Exam Questions

Q1 Describe the tests you would carry out in order to distinguish between solid samples of sodium chloride and sodium bromide using:

a) silver nitrate solution and aqueous ammonia,

b) concentrated sulfuric acid.

For each test, state your observations and write equations for the reactions which occur.   [11 marks]

Q2 The halogen below iodine in Group 7 is astatine (At). Predict, giving an explanation, whether or not:

a) hydrogen sulfide gas would be evolved when concentrated sulfuric acid is added to a solid
   sample of sodium astatide.   [4 marks]

b) silver astatide will dissolve in concentrated ammonia solution.   [3 marks]

[Sing along with me] "Why won't this section end... Why won't this section end..."

AS Chemistry. What a bummer, eh... No one ever said it was going to be easy. Not even your teacher would be that cruel. There are plenty more equations on this page to learn. As well as that, make sure you really understand everything... the trend in the reducing power of halides... and the reactions with sulfuric acid... And no, you can't swap to English. Sorry.

Group 2 — The Alkaline Earth Metals

Group 2, AKA the alkaline earth metals, are in the "s block" of the periodic table. There's three pages about these jolly fellas and their compounds, so we've got a lot to do — best get on...

1) Group 2 Elements **Lose Two Electrons** when they React

Element	Atom	Ion
Be	$1s^2 2s^2$	$1s^2$
Mg	$1s^2 2s^2 2p^6 3s^2$	$1s^2 2s^2 2p^6$
Ca	$1s^2 2s^2 2p^6 3s^2 3p^6 4s^2$	$1s^2 2s^1 2p^6 3s^2 3p^6$

Group 2 elements all have two electrons in their outer shell (s^2).

They lose their two outer electrons to form **2+ ions**. Their ions then have every atom's dream electronic structure — that of a **noble gas**.

2) Atomic Radius **Increases** Down the Group

This is because of the extra **electron shells** as you go down the group.

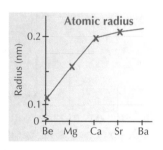

3) Ionisation Energy **Decreases** Down the Group

1) Each element down Group 2 has an **extra electron shell** compared to the one above.

2) The extra inner shells **shield** the outer electrons from the attraction of the nucleus.

3) Also, the extra shell means that the outer electrons are **further away** from the nucleus, which greatly reduces the nucleus's attraction.

Mr Kelly has one final attempt at explaining electron shielding to his students...

Both of these factors make it **easier** to remove outer electrons, resulting in a **lower ionisation energy**.

The positive charge of the nucleus does increase as you go down a group (due to the extra protons), but this effect is overridden by the effect of the extra shells.

4) Reactivity **Increases** Down the Group

1) As you go down the group, the **ionisation energies** decrease. This is due to the increasing atomic radius and shielding effect (see above).

2) When Group 2 elements react they **lose electrons**, forming positive ions. The easier it is to lose electrons (i.e. the lower the first and second ionisation energies), the more reactive the element, so **reactivity increases** down the group.

5) Melting Points Generally **Decrease** Down the Group

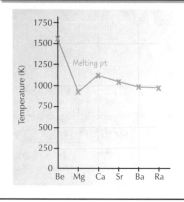

1) The Group 2 elements have typical **metallic structures**, with the electrons of their outer shells being **delocalised**.

2) Going down the group the metallic ions get **bigger** — so they have a smaller **charge/volume ratio**. But the number of delocalised electrons per atom doesn't change (it's always 2) — so the delocalised electrons get more **spread out**.

3) These two factors mean there's reduced attraction of the positive ions to the 'sea' of delocalised electrons. So it takes **less energy** to break the bonds, which means lower melting points generally down the group. However, there's a big 'blip' at magnesium, because the crystal structure (the arrangement of the metallic ions) changes.

Group 2 — The Alkaline Earth Metals

Group 2 Elements React With *Water*

When Group 2 elements react, they are **oxidised** from a state of **0** to **+2**, forming M^{2+} ions.

$$M \rightarrow M^{2+} + 2e^-$$

Oxidation state: 0   +2   E.g. $Ca \rightarrow Ca^{2+} + 2e^-$

                                      0   +2

The Group 2 metals react with water to give a **metal hydroxide and hydrogen**.

$$M_{(s)} + 2H_2O_{(l)} \rightarrow M(OH)_{2\,(aq)} + H_{2\,(g)}$$

Oxidation state: 0   +2

e.g. $Ca_{(s)} + 2H_2O_{(l)} \rightarrow Ca(OH)_{2\,(aq)} + H_{2\,(g)}$

They react **more readily** down the group because the **ionisation energies** decrease.

Be	doesn't react
Mg	VERY slowly
Ca	steadily
Sr	fairly quickly
Ba	rapidly

Right, that's enough about the Group 2 elements. From here on, we're looking at their cuddly compounds...

(Sorry, just trying to liven things up a bit.)

Solubility Trends Depend on the *Compound Anion*

Generally, compounds of Group 2 elements that contain **singly charged** negative ions (e.g. OH^-) **increase** in solubility down the group, whereas compounds that contain **doubly charged** negative ions (e.g. SO_4^{2-}) **decrease** in solubility down the group.

Group 2 element	hydroxide (OH^-)	sulfate (SO_4^{2-})
magnesium	least soluble	most soluble
calcium		
strontium		
barium	most soluble	least soluble

Compounds like magnesium hydroxide which have **very low** solubilities are said to be **sparingly soluble**.

Most sulfates are soluble in water, but **barium sulfate** is **insoluble**.

The test for sulfate ions makes use of this property...

Test for sulfate ions

If acidified barium chloride ($BaCl_2$) is added to a solution containing sulfate ions then a white precipitate of barium sulfate is formed.

$$Ba^{2+}_{(aq)} + SO_4^{2-}_{(aq)} \rightarrow BaSO_{4\,(s)}$$

You need to acidify the barium chloride (with, say, hydrochloric acid) to get rid of any lurking sulfites or carbonates.

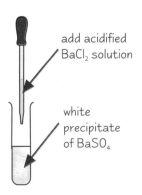

add acidified $BaCl_2$ solution

white precipitate of $BaSO_4$

Group 2 — The Alkaline Earth Metals

Group 2 Compounds are used to Neutralise Acidity

Group 2 elements are known as the **alkaline earth metals**, and many of their common compounds are used for neutralising acids. Here are a couple of common examples:

1) Calcium hydroxide (slaked lime, $Ca(OH)_2$) is used in **agriculture** to neutralise acid soils.

2) Magnesium hydroxide ($Mg(OH)_2$) is used in some indigestion tablets as an **antacid**.

In both cases, the ionic equation for the neutralisation is
$$H^+_{(aq)} + OH^-_{(aq)} \rightarrow H_2O_{(l)}$$

Daisy the cow *

Barium Sulfate is Used in 'Barium Meals'

CHRIS PRIEST / SCIENCE PHOTO LIBRARY

X-rays are great for finding broken bones, but they pass straight through soft tissue — so soft tissues, like the digestive system, don't show up on conventional X-ray pictures.

1) Barium sulfate is **opaque** to X-rays — they won't pass through it. It's used in **'barium meals'** to help diagnose problems with the oesophagus, stomach or intestines.

2) A patient swallows the barium meal, which is a suspension of **barium sulfate**. The barium sulfate **coats** the tissues, making them show up on the X-rays, showing the structure of the organs.

Practice Questions

Q1 Which is the least reactive metal in Group 2? Why does reactivity with water increase down Group 2?

Q2 Which of the following increases in size down Group 2? **atomic radius, first ionisation energy, boiling point**

Q3 Which is less soluble, barium sulfate or magnesium sulfate?

Q4 How is the solubility of magnesium hydroxide often described?

Q5 Give a use of magnesium hydroxide.

Exam Questions

Q1 Use the electron configurations of magnesium and calcium to help explain the difference between their first ionisation energies. [5 marks]

Q2 The table shows the atomic radii of three elements from Group 2.

Element	Atomic radius/nm
X	0.089
Y	0.198
Z	0.176

a) Predict which element would react most rapidly with water. [1 mark]

b) Explain your answer. [2 marks]

Q2 Describe how you could use barium chloride solution to distinguish between solutions of zinc chloride and zinc sulfate. Give the expected observations and an appropriate balanced equation including state symbols. [4 marks]

Q4 Hydrochloric acid can be produced in excess quantities in the stomach, causing indigestion. Antacid tablets often contain sodium hydrogencarbonate ($NaHCO_3$), which reacts with the acid to form a salt, carbon dioxide and water.

a) Write an equation for the neutralisation of hydrochloric acid with sodium hydrogencarbonate. [1 mark]

b) What discomfort could be caused by the carbon dioxide produced? [1 mark]

c) From your knowledge of Group 2 compounds, choose an alternative antacid that would not give this problem and write an equation for its reaction with hydrochloric acid. [2 marks]

Bored of Group 2 trends? Me too. Let's play noughts and crosses...

Noughts and crosses is pretty rubbish really, isn't it?
It's always a draw. Ho hum. Back to Chemistry then, I guess...

* She wanted to be in the book. I said OK.

Extraction of Metals

Metals are handy for making metal things. Sadly, you don't just find big lumps of pure metal lying about, ready to use...

Sulfide Ores are Usually Converted to Oxides First

1) An **ore** is a natural substance that a **metal** can be economically extracted from. In other words, a rock you can get quite a bit of metal out of.

2) Metals are often found in ores as **sulfides** (such as lead sulfide and zinc sulfide), or **oxides** (like titanium dioxide and iron(III) oxide). The metal **element** needs to be removed from these compounds — that's where the chemistry comes in.

3) The first step to extract a metal from a **sulfide ore** is to turn it into an **oxide**. This is done by **roasting** the sulfide in air.

When extracting metals, Jimmy liked to use his ores and cart.

E.g.

$$\text{zinc sulfide} + \text{oxygen} \rightarrow \text{zinc oxide} + \text{sulfur dioxide}$$
$$2ZnS_{(s)} + 3O_{2(g)} \rightarrow 2ZnO_{(s)} + 2SO_{2(g)}$$

4) Here's the bad news: **sulfur dioxide** gas causes **acid rain**. Acid rain can cause harm to plants and aquatic life, and damage limestone buildings, so the sulfur dioxide can't be **released** into the atmosphere.

5) But here's the good news: by converting the sulfur dioxide to **sulfuric acid** a pollutant is avoided, and a **valuable product** is made — sulfuric acid's in demand because it's used in many chemical and manufacturing processes.

Oxides are Reduced to the Metal

1) The method for **reducing** the oxide depends on the metal you're trying to extract.

2) **Carbon** (as coke — a solid fuel made from coal) and **carbon monoxide** are used as **reducing agents** for quite a few metals — usually the ones that are **less reactive** than carbon.

You need to know these three **examples** of the extraction of metals with carbon and carbon monoxide:

REDUCTION OF IRON(III) OXIDE
Iron(III) oxide is reduced by **carbon** or **carbon monoxide** to iron and carbon dioxide.

$$2Fe_2O_3 + 3C \rightarrow 4Fe + 3CO_2$$
$$Fe_2O_3 + 3CO \rightarrow 2Fe + 3CO_2$$

This happens in a blast furnace at temperatures greater than 700 °C.

REDUCTION OF MANGANESE(IV) OXIDE (MANGANESE DIOXIDE)
Manganese(IV) oxide is reduced with **carbon** (as coke) or **carbon monoxide** in a blast furnace.

$$MnO_2 + C \rightarrow Mn + CO_2$$
$$MnO_2 + 2CO \rightarrow Mn + 2CO_2$$

This needs higher temperatures than iron(III) oxide — about 1200 °C.

REDUCTION OF COPPER CARBONATE
Copper can be extracted using **carbon**.
One ore of copper is **malachite**, containing $CuCO_3$. This can be heated directly with carbon.

$$2CuCO_3 + C \rightarrow 2Cu + 3CO_2$$

Another method involves heating the carbonate until it decomposes, then reducing the oxide with carbon.

$$CuCO_3 \rightarrow CuO + CO_2$$
$$2CuO + C \rightarrow 2Cu + CO_2$$

Carbon and **carbon monoxide** are the first choice for extracting metals because they're **cheap**. But they're not always suitable — some metals have to be extracted by other methods. You'll see three examples on the next page...

Extraction of Metals

Tungsten is Extracted Using Hydrogen

1) **Tungsten** can be extracted from its oxide with carbon, but that can leave **impurities** which make the metal more **brittle**. If pure tungsten is needed, the ore is reduced using **hydrogen** instead.

$$WO_{3(s)} + 3H_{2(g)} \rightarrow W_{(s)} + 3H_2O_{(g)}$$

This happens in a furnace at temperatures above 700 °C.

2) Tungsten is the **only metal** reduced on a large scale using **hydrogen**. Hydrogen is more **expensive** but it's worth the extra cost to get pure tungsten, which is much easier to work with.

3) Hydrogen is **highly explosive** when mixed with air though, which is a bit of a hazard.

Aluminium is Extracted by Electrolysis

1) **Aluminium** is **too reactive** to extract using reduction by carbon. A very **high temperature** is needed, so extracting aluminium by reduction is **too expensive** to make it worthwhile.

2) Aluminium's ore is called **bauxite** — it's aluminium oxide, Al_2O_3, with various impurities. First of all, these impurities are removed. Next, it's dissolved in **molten cryolite** (sodium aluminium fluoride, Na_3AlF_6), which lowers its **melting point** from a scorching 2050 °C, to a cool **970 °C**. This reduces the operating costs.

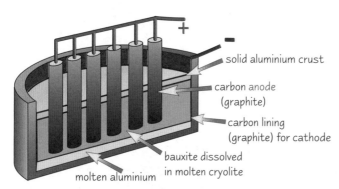

- solid aluminium crust
- carbon anode (graphite)
- carbon lining (graphite) for cathode
- bauxite dissolved in molten cryolite
- molten aluminium

The current used in electrolysis is high (200 000 A), so the process is carried out where cheap electricity is available, often near hydroelectric power stations.

ELECTROLYSIS OF ALUMINIUM

1) Aluminium is produced at the **cathode** and collects as the molten liquid at the bottom of the cell.

$$Al^{3+} + 3e^- \rightarrow Al$$

2) Oxygen is produced at the **anode**.

$$2O^{2-} \rightarrow O_2 + 4e^-$$

Titanium is used in the bodies of modern planes.

Titanium is Great but a bit too Expensive

1) Titanium is a pretty **abundant** metal in the Earth's crust. In its pure form, titanium is a **strong, light** metal that is highly resistant to **corrosion**. Pretty much perfect really, so how come it's not used more... Well basically, it's just a bit too **difficult** and **expensive** to produce.

2) The main ore is rutile (titanium(IV) oxide, TiO_2). You can't extract titanium from it by carbon reduction because you get titanium carbide which ruins it... $TiO_{2\,(s)} + 3C_{(s)} \rightarrow TiC_{(s)} + 2CO_{(g)}$

THE EXTRACTION OF TITANIUM

...is a **batch** process with several stages.

1) The ore is converted to **titanium(IV) chloride** by heating it to about 900 °C with carbon in a stream of chlorine gas.

$$TiO_{2\,(s)} + 2Cl_{2\,(g)} + 2C_{(s)} \rightarrow TiCl_{4\,(g)} + 2CO_{(g)}$$

2) The titanium chloride is purified by **fractional distillation** under an inert atmosphere of argon or nitrogen.

3) Then the chloride gets reduced in a **furnace** at almost 1000 °C. It's heated with a **more reactive** metal such as sodium or magnesium. An inert atmosphere is used to prevent side reactions.

$$TiCl_{4\,(g)} + 4Na_{(l)} \rightarrow Ti_{(s)} + 4NaCl_{(l)}$$
$$TiCl_{4\,(g)} + 2Mg_{(l)} \rightarrow Ti_{(s)} + 2MgCl_{2\,(l)}$$

Na and Mg are reducing agents.

Extraction of Metals

Recycling can be Good for the Environment and Save Money

Once you've got the metal out of the ore, you can keep **recycling** it again and again.
As usual, there are pros and cons:

Advantages of recycling metals:

- Saves raw materials — ores are a finite resource.
- Saves energy — recycling metals takes less energy than extracting metal. This saves money too.
- Reduces waste sent to landfill.
- Mining damages the landscape and spoil heaps are ugly. Recycling metals reduces this.

Disadvantages of recycling metals:

- Collecting and sorting metals from other waste can be difficult and expensive.
- The purity of recycled metal varies — there's usually other metals and other impurities mixed in.
- Recycling metals may not produce a consistent supply to meet demand.

Scrap Iron can be used in Copper Extraction

1) Some scrap metal can be put to other uses. For example, **scrap iron** can be used to extract **copper** from solution. This method is mainly used with **low grade** ore — ore that only contains a **small percentage** of copper.

2) Acidified water **dissolves** the copper compounds in the ore.
The solution is collected and **scrap iron** is then added. The iron dissolves and **reduces** the copper(II) ions.
The copper precipitates out of the solution. $Cu^{2+}_{(aq)} + Fe_{(s)} \rightarrow Cu_{(s)} + Fe^{2+}_{(aq)}$

3) This process produces copper **more slowly** than carbon reduction and has a **lower yield**, which is why it's not used with ores that have a high copper content. It's **cheaper** than carbon reduction though, because you don't need **high temperatures**, and better for the environment because there's no CO_2 produced.

Practice Questions

Q1 What is used to reduce manganese(IV) oxide to manganese?

Q2 Write the equation for the conversion of zinc sulfide to zinc oxide.

Q3 Write the equation for the displacement of titanium from titanium chloride using sodium.

Q4 Why is sodium (or magnesium) chosen to reduce titanium chloride?

Q5 Give one environmental and one economic reason for recycling metals.

Exam Questions

Q1 The iron ore in a blast furnace contains a mixture of oxides, one of which is Fe_3O_4.
When Fe_3O_4 is reduced, both carbon and carbon monoxide act as reducing agents.
Write equations to show
a) how carbon monoxide reduces Fe_3O_4. [1 mark]
b) how carbon reduces Fe_3O_4. [1 mark]

Q2 Hydrogen is used to extract tungsten from its ore.
Give an advantage and a disadvantage of using hydrogen in place of carbon. [2 marks]

Q3 Aluminium is extracted from its purified ore by electrolysis.
a) What important step is taken to reduce the cost of extracting aluminium? [2 marks]
b) Write equations for the reactions occurring at each electrode. [2 marks]
c) Explain why aluminium is more expensive to extract than iron. [1 mark]

Extraction can be heavy going — in fact, it's like pulling teeth...

It might look like there are loads of equations to learn here — in fact, come to think of it, there are quite a few. But at least most of those oxide reduction ones are pretty similar, e.g. oxide + carbon monoxide → metal + carbon dioxide. There's actually <u>nothing too hard</u> on these pages at all. But there is <u>plenty</u> to get stuck into with your revision shovel.

Synthesis of Chloroalkanes

Wow, you've reached the last section of the book already — time flies when you're having fun...

Alkanes **Don't React** with Most Chemicals

1) The C–C bonds and C–H bonds in alkanes are pretty **non-polar**. But most chemicals are **polar** — like water, haloalkanes, acids and alkalis.

2) Polar chemicals are attracted to the **polar groups** on molecules they attack. Alkanes don't have any polar groups, so they **don't** react with polar chemicals.

3) Alkanes **will** react with some **non-polar** things though — such as oxygen or the halogens. But they'll **only** bother if you give them enough **energy**.

If all this polar talk means nothing to you, flick back to p28. It's all explained there.

Halogens React with **Alkanes**, Forming **Haloalkanes**

(P68 has more detail on haloalkanes.)

1) Halogens react with alkanes in **photochemical** reactions. Photochemical reactions are started by **ultraviolet** light.

2) A hydrogen atom is **substituted** (replaced) by chlorine or bromine. This is a **free-radical substitution reaction**.

Free radicals are particles with an unpaired electron, written like this — $Cl\cdot$ or $CH_3\cdot$. You get them when bonds split equally, and they're highly reactive.

Chlorine and **methane** react with a bit of a bang to form **chloromethane**:

$$CH_4 + Cl_2 \xrightarrow{U.V.} CH_3Cl + HCl$$

The **reaction mechanism** has three stages:

Initiation reactions — free radicals are produced.

1) Sunlight provides enough energy to break the Cl-Cl bond — this is **photodissociation**.

$$Cl_2 \xrightarrow{U.V.} 2Cl\cdot$$

2) The bond splits **equally** and each atom gets to keep one electron. The atom becomes a highly reactive **free radical**, $Cl\cdot$, because of its **unpaired electron**.

Propagation reactions — free radicals are used up and created in a chain reaction.

1) $Cl\cdot$ attacks a **methane** molecule: $Cl\cdot + CH_4 \rightarrow CH_3\cdot + HCl$

2) The new **methyl free radical**, $CH_3\cdot$, can attack another Cl_2 molecule: $CH_3\cdot + Cl_2 \rightarrow CH_3Cl + Cl\cdot$

3) The new $Cl\cdot$ can attack **another** CH_4 molecule, and so on, until all the Cl_2 or CH_4 molecules are wiped out.

Termination reactions — free radicals are mopped up.

1) If two free radicals join together, they make a **stable molecule**.

2) There are **heaps** of possible termination reactions. Here's a couple of them to give you the idea: $Cl\cdot + CH_3\cdot \rightarrow CH_3Cl$

$$CH_3\cdot + CH_3\cdot \rightarrow C_2H_6$$

Some products formed will be trace impurities in the final sample.

More substitutions

What happens now **depends** on whether there's too much **chlorine** or too much **methane**:

1) If the **chlorine's** in excess, $Cl\cdot$ free radicals will start attacking chloromethane, producing **dichloromethane** CH_2Cl_2, **trichloromethane** $CHCl_3$, and **tetrachloromethane** CCl_4.

2) **But** if the **methane's** in excess, then the product will mostly be **chloromethane**.

Chloroalkanes and **Chlorofluoroalkanes** are Used as **Solvents**

1) **Chlorofluorocarbons** (**CFCs**) are haloalkane molecules where all of the hydrogen atoms have been replaced by **chlorine** and **fluorine** atoms. E.g.

2) Both **CFCs** and **chloroalkanes** can be used as **solvents** — they both used to be used in dry cleaning and degreasing.

trichlorofluoromethane   chlorotrifluoromethane

Synthesis of Chloroalkanes

Chlorine Atoms are Destroying The Ozone Layer

1) Ozone in the upper atmosphere acts as a **chemical sunscreen**. It absorbs a lot of the **ultraviolet radiation** which can cause sunburn or even skin cancer.

2) Ozone's **formed naturally** when an **oxygen molecule** is **broken down** into **two free radicals** by **ultraviolet radiation**. The free radicals **attack** other oxygen molecules forming **ozone**.
Just like this:

$$O_2 + h\nu \rightarrow O\bullet + O\bullet \Longrightarrow O_2 + O\bullet \rightarrow O_3$$

a quantum of UV radiation

You've heard of how the **ozone layer's** being destroyed by **CFCs**, right. Well, here's what's happening.

1) **Chlorine free radicals**, $Cl\bullet$, are formed when the C–Cl bonds in **CFCs** are broken down by **ultraviolet radiation**.

E.g. $CCl_3F_{(g)} \rightarrow CCl_2F\bullet_{(g)} + Cl\bullet_{(g)}$

2) These free radicals are **catalysts**. They react with **ozone** to form an **intermediate** ($ClO\bullet$), and an **oxygen molecule**.

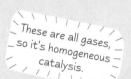
These are all gases, so it's homogeneous catalysis.

$$Cl\bullet_{(g)} + O_{3(g)} \rightarrow O_{2(g)} + ClO\bullet_{(g)}$$
$$ClO\bullet_{(g)} + O_{3(g)} \rightarrow 2O_{2(g)} + Cl\bullet_{(g)}$$

The chlorine free radical is regenerated. It goes straight on to attack another ozone molecule. It only takes one little chlorine free radical to destroy loads of ozone molecules.

3) So the **overall reaction is...** $2O_{3(g)} \rightarrow 3O_{2(g)}$ ... and $Cl\bullet$ is the catalyst.

CFCs Are Now Banned

1) CFCs are pretty **unreactive**, **non-flammable** and **non-toxic**. They used to be used in fire extinguishers, as propellants in aerosols, as the coolant gas in fridges and to foam plastics to make insulation and packaging materials.

2) In the 1970s scientists discovered that CFCs were causing **damage** to the **ozone layer**.
The **advantages** of CFCs couldn't outweigh the **environmental problems** they were causing, so they were **banned**.

3) Chemists have developed **alternatives** to CFCs. **HCFCs (hydrochlorofluorocarbons)** and **HFCs (hydrofluorocarbons)** are less dangerous than CFCs, so they're being used as temporary alternatives until safer products are developed.

4) Most aerosols now have been replaced by **pump spray systems** or use **nitrogen** as the propellant. Many industrial fridges use **ammonia** or **hydrocarbons** as the coolant gas, and **carbon dioxide** is used to make foamed polymers.

Practice Questions

Q1 What's a free radical?

Q2 What's photodissociation?

Q3 Give two uses of CFCs.

Q4 Describe how ozone is beneficial.

Q5 Write an equation to show what happens when UV radiation breaks down CFCs.

Q6 What is the formula for ozone?

Q7 What products are formed when a chlorine free radical reacts with an ozone molecule?

Exam Question

Q1 The alkane ethane is a saturated hydrocarbon. It is mostly unreactive, but will react with chlorine in a photochemical reaction.

(a) What is a saturated hydrocarbon? [2 marks]

(b) Why is ethane unreactive with most reagents? [2 marks]

(c) Write an equation and outline the mechanism for the photochemical reaction of chlorine with ethane. Assume ethane is in excess. What type of mechanism is it? [8 marks]

This stuff is like...totally radical, man...

Mechanisms are an absolute pain in the bum to learn, but unfortunately reactions are what Chemistry's all about. If you don't like it, you should have taken art — no mechanisms in that, just pretty pictures. Ah well, there's no going back now. You've just got to sit down and learn the stuff. Keep hacking away at it, till you know it all off by heart.

Haloalkanes

Don't worry if you see haloalkanes called halogenoalkanes. It's a government conspiracy to confuse you.

Haloalkanes are Alkanes with Halogen Atoms

A **haloalkane** is an alkane with at least one **halogen atom** in place of a hydrogen atom.

E.g.

| trichloromethane | 2-iodopropane | 2-bromo-2-chloro-1, 1, 1-trifluoroethane |

The Carbon–Halogen Bond in Haloalkanes is Polar

1) Halogens are much more **electronegative** than carbon.
   So, the **carbon-halogen bond** is polar.

2) The **δ+ carbon** doesn't have enough electrons.
   This means it can be attacked by a **nucleophile**.
   A nucleophile's an **electron-pair donor**.
   It donates an electron pair to somewhere without enough electrons.

3) **OH⁻, CN⁻** and **NH₃** are all **nucleophiles** which react with haloalkanes.

$$-\overset{|}{\underset{|}{C}}-Br^{\delta-}$$
$$\delta+$$

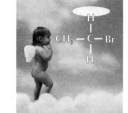

Haloalkanes are special
amongst alkanes...

Haloalkanes can be Hydrolysed to make Alcohols

For example, bromoethane can be hydrolysed to ethanol. You have to use **warm aqueous sodium** or **potassium hydroxide** or it won't work. It's a **nucleophilic substitution reaction**.

Here's how it happens:

Curly arrows (and that's an official term) show the movement of an electron pair.

The OH⁻ ion acts as a nucleophile, attacking the slightly positive carbon atom.

The C-Br bond is polar. The C$^{\delta+}$ attracts a lone pair of electrons from the OH⁻ ion.

The C-Br bond breaks and both the electrons are taken by the Br. A new bond forms between the C and the OH⁻ ion.

Here's the general equation for this reaction: **R–X + OH⁻ → ROH + X⁻**

R represents an alkyl group. X stands for one of the halogens (F, Cl, Br or I).

Iodoalkanes are Hydrolysed the Fastest

1) The **carbon-halogen bond strength** (or enthalpy) decides **reactivity**.
   For any reaction to occur the carbon-halogen bond needs to **break**.

2) The **C-F bond** is the **strongest** — it has the highest **bond enthalpy**.
   So **fluoroalkanes** are hydrolysed **more slowly** than other haloalkanes.

3) The **C-I bond** has the **lowest bond enthalpy**, so it's easier to break.
   This means that **iodoalkanes** are hydrolysed more **quickly**.

bond	bond enthalpy kJ mol⁻¹
C–F	467
C–Cl	346
C–Br	290
C–I	228

Faster hydrolysis as bond enthalpy decreases (the bonds are getting weaker).

Haloalkanes

Haloalkanes React With Ammonia to Form Amines

If you **warm** a haloalkane with excess **ethanolic** ammonia, the **ammonia** swaps places with the **halogen** — yes, it's another one of those **nucleophilic substitution reactions**.

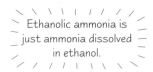
Ethanolic ammonia is just ammonia dissolved in ethanol.

The first step is the same as in the mechanism on the previous page, except this time the nucleophile is NH_3.

(See page 78 to find out what reflux means.)

In the second step, an ammonia molecule removes a hydrogen from the NH_3 group to form an ammonium ion (NH_4^+).

The ammonium ion can react with the bromine ion to form ammonium bromide. So the overall reaction is this:

$$CH_3-\overset{\overset{\displaystyle H}{|}}{\underset{\underset{\displaystyle H}{|}}{C}}-Br \; + \; 2NH_3 \; \xrightarrow[\text{ethanol}]{\text{reflux}} \; CH_3-\overset{\overset{\displaystyle H}{|}}{\underset{\underset{\displaystyle H}{|}}{C}}-NH_2 \; + \; NH_4Br$$

You can use Haloalkanes to form Nitriles

Nitriles have $-C\equiv N$ groups.

If you **warm** a haloalkane with **ethanolic potassium cyanide** you get a **nitrile**.

It's yet another **nucleophilic substitution reaction** — the **cyanide ion**, CN^-, is the **nucleophile**.

Haloalkanes also Undergo Elimination Reactions

If you warm a haloalkane with hydroxide ions dissolved in **ethanol** instead of water, an **elimination reaction** happens, and you end up with an **alkene**. This is how you do it:

1) Heat the mixture **under reflux** or you'll lose volatile stuff.

$$CH_3CHBrCH_3 + KOH \xrightarrow[\text{reflux}]{\text{ethanol}} CH_2{=}CH_2CH_3 + H_2O + KBr$$

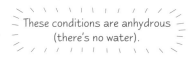
These conditions are anhydrous (there's no water).

2) Here's how the reaction works:

OH^- acts as a base and takes a proton, H^+, from the carbon on the left. This makes water. The left carbon now has a spare electron, so it forms a double bond with the other carbon. To form the double bond, the right carbon has to let go of the Br, which drops off as a Br^- ion.

3) This is an example of an **elimination reaction**. In an elimination **reaction**, a **small group** of atoms breaks away from a larger molecule. This **small group** is **not replaced** by anything else (whereas it would be in a substitution reaction).

In the reaction above, H and Br have been eliminated from CH_3CH_2Br to leave $CH_2{=}CH_2$

Haloalkanes

The Type of Reaction Depends on the Conditions

You can control what **type of reaction** happens by **changing the conditions**.

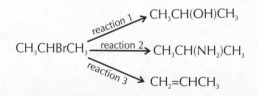

Aqueous conditions —
nucleophilic substitution

OH^- acts as a nucleophile

Anhydrous conditions —
elimination

OH^- acts as a base

Both of these reactions have their uses...

1) The **elimination** reaction is a **good way** of getting a **double bond** into a molecule. Loads of other organic synthesis reactions use **alkenes**, so the elimination reaction is a good starting point for making lots of different organic chemicals.

2) The **substitution** reaction allows you to produce any **alcohol** molecule that you need. And alcohols can be the starting point for synthesis reactions that produce **aldehydes**, **ketones**, **esters**, and **carboxylic acids**.

So haloalkanes are very useful as a **starting material** for making other organic compounds.

Practice Questions

Q1 What is a haloalkane?

Q2 What is a nucleophile?

Q3 Why is the carbon-halogen bond polar?

Q4 Why does iodoethane react faster than chloro- or bromoethane with warm, aqueous sodium hydroxide?

Exam Question

Q1 Three reactions of 2-bromopropane, $CH_3CHBrCH_3$, are shown below.

$$CH_3CHBrCH_3 \xrightarrow{\text{reaction 1}} CH_3CH(OH)CH_3$$
$$\xrightarrow{\text{reaction 2}} CH_3CH(NH_2)CH_3$$
$$\xrightarrow{\text{reaction 3}} CH_2=CHCH_3$$

a) For each reaction, name the reagent and solvent used.   [6 marks]

b) Under the same conditions, 2-iodopropane was used in reaction 1 in place of 2-bromopropane. What difference (if any) would you expect in the rate of the reaction? Explain your answer.   [2 marks]

If you don't learn this — you will be eliminated. Resistance is nitrile...

Polar bonds get in just about every area of Chemistry. If you still think they're something to do with either bears or mints, flick back to page 28 and have a good read. Make sure you learn these reactions, and the mechanisms, as well as which bonds are hydrolysed fastest. This stuff's always coming up in exams. Ruin the examiner's day and get it right.

Reactions of Alkenes

I'll warn you now — some of this stuff gets a bit heavy — but stick with it, as it's pretty important.

Alkenes are Unsaturated Hydrocarbons

1) Alkenes have the **general formula C_nH_{2n}**. They're just made of carbon and hydrogen atoms, so they're **hydrocarbons**.
2) Alkene molecules **all** have at least one **C=C double covalent bond**. Molecules with C=C double bonds are **unsaturated** because they can make more bonds with extra atoms in **addition** reactions.
3) Because there's two pairs of electrons in the C=C double bond, it has a really **high electron density**. This makes alkenes pretty reactive.

Here are a few pretty diagrams of **alkenes**:

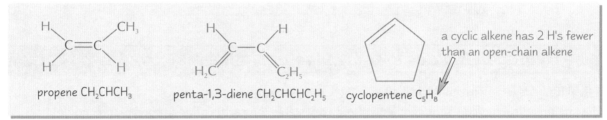

propene CH_2CHCH_3 penta-1,3-diene $CH_2CHCHC_2H_5$ cyclopentene C_5H_8

a cyclic alkene has 2 H's fewer than an open-chain alkene

Electrophilic Addition Reactions Happen to Alkenes

Electrophilic addition reactions aren't too complicated...

1) The **double bonds** open up and atoms are **added** to the carbon atoms.
2) Electrophilic addition reactions happen because the double bond has got plenty of **electrons** and is easily attacked by **electrophiles**.

> **Electrophiles** are **electron-pair acceptors** — they're usually a bit short of electrons, so they're <u>attracted</u> to areas where there's lots of them about.
>
> Here's a few examples:
> - **Positively charged ions**, like H^+, NO_2^+.
> - **Polar molecules** — the δ+ atom is attracted to places with lots of electrons

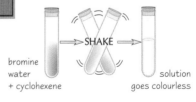

See page 28 for a reminder about polar molecules.

3) The double bond is also **nucleophilic** — it's attracted to places that don't have enough **electrons**.

Use Bromine Water to Test for C=C Double Bonds

When you shake an alkene with **orange bromine water**, the solution quickly **decolourises**. Bromine is added across the double bond to form a colourless **dibromoalkane** — this happens by **electrophilic addition**. Here's the mechanism...

bromine water + cyclohexene — SHAKE → solution goes colourless

$$H_2C=CH_2 + Br_2 \rightarrow CH_2BrCH_2Br$$

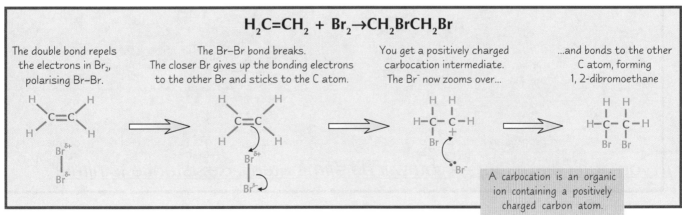

The double bond repels the electrons in Br_2, polarising Br–Br. | The Br–Br bond breaks. The closer Br gives up the bonding electrons to the other Br and sticks to the C atom. | You get a positively charged carbocation intermediate. The Br⁻ now zooms over... | ...and bonds to the other C atom, forming 1,2-dibromoethane

A carbocation is an organic ion containing a positively charged carbon atom.

Reactions of Alkenes

Alkenes also Undergo **Addition** with **Hydrogen Halides**

Alkenes also undergo **addition** reactions with hydrogen bromide — to form **bromoalkanes**.
This is the reaction between **ethene** and HBr:

$$H_2C=CH_2 + HBr \longrightarrow CH_3CH_2Br$$

(Other alkenes react in a similar way.)

$$C_2H_4 + HBr \rightarrow C_2H_5Br$$

Adding **Hydrogen Halides** to **Unsymmetrical Alkenes** Forms **Two Products**

1) If the HBr adds to an **unsymmetrical** alkene, there are two possible products.

2) The amount of each product formed depends on how **stable** the **carbocation** formed in the middle of the reaction is.

3) Carbocations with more **alkyl groups** are more stable because the alkyl groups feed **electrons** towards the positive charge. The **more stable carbocation** is much more likely to form.

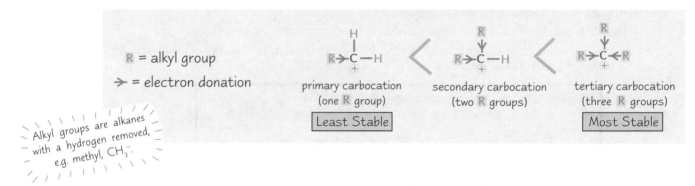

R = alkyl group

→ = electron donation

primary carbocation
(one R group)
Least Stable

secondary carbocation
(two R groups)

tertiary carbocation
(three R groups)
Most Stable

Alkyl groups are alkanes with a hydrogen removed, e.g. methyl, CH_3^-.

Here's how hydrogen bromide reacts with propene:

$$H_2C=CHCH_3 + HBr \longrightarrow CH_3CHBrCH_3$$
2-bromopropane
(major product)

$$H_2C=CHCH_3 + HBr \longrightarrow CH_2BrCH_2CH_3$$
1-bromopropane
(minor product)

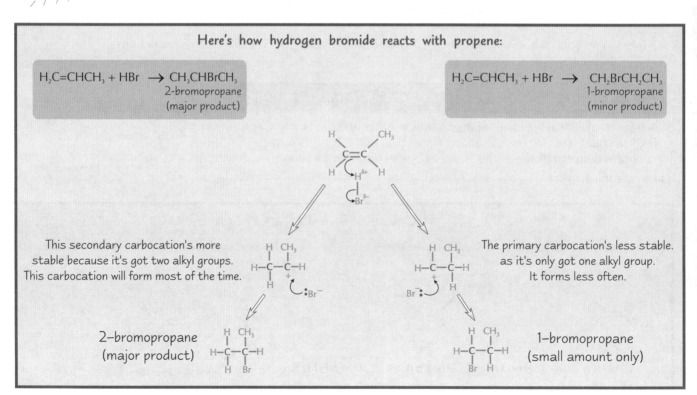

This secondary carbocation's more stable because it's got two alkyl groups. This carbocation will form most of the time.

The primary carbocation's less stable, as it's only got one alkyl group. It forms less often.

2-bromopropane
(major product)

1-bromopropane
(small amount only)

Reactions of Alkenes

Reacting Alkenes with *Water* and an *H₂SO₄* Catalyst Makes *Alcohols*

Alcohols are produced industrially by **hydrating alkenes** in the presence of an **acid catalyst**, such as sulfuric acid:

1) Cold concentrated **sulfuric acid** reacts with an alkene in an **electrophilic addition** reaction.

$$H_2C = CH_2 + H_2SO_4 \longrightarrow CH_3CH_2OSO_2OH$$
ethene   sulfuric acid   ethyl hydrogen sulfate

2) If you then add cold **water** and warm the product, it's **hydrolysed** to form an alcohol.

$$CH_3CH_2OSO_2OH + H_2O \longrightarrow CH_3CH_2OH + H_2SO_4$$
ethyl hydrogen sulfate   ethanol

3) The **sulfuric acid** isn't used up — it acts as a **catalyst**.

> Hydrolysis is the breaking of covalent bonds by reaction with water.

So the overall reaction is: $H_2C = CH_2 + H_2O \xrightarrow{H_2SO_4} C_2H_5OH$

Ethanol is Manufactured by *Steam Hydration*

1) Ethene can be **hydrated** by **steam** at 300 °C and a pressure of 60 atm. It needs a solid **phosphoric(V) acid catalyst**.

2) The reaction's **reversible** and the reaction yield is low — only about 5%. This sounds rubbish, but you can **recycle** the unreacted ethene gas, making the overall yield a much more profitable **95%**.

$$H_2C=CH_{2(g)} + H_2O_{(g)} \underset{\substack{300\,°C \\ 60\,atm}}{\overset{H_3PO_4}{\rightleftharpoons}} CH_3CH_2OH_{(g)}$$

Practice Questions

Q1 What's the general formula for an alkene?

Q2 What is an electrophile?

Q3 Why do alkenes react with electrophiles?

Q4 What is a carbocation?

Q5 What conditions are needed to produce ethanol from ethene?

Exam Question

Q1 But-1-ene is an alkene. Alkenes contain at least one C=C double bond.
   a) Describe how bromine water can be used to test for C=C double bonds.   [2 marks]

   b) Name the reaction mechanism involved in the above test.   [2 marks]

   c) Hydrogen bromide will react with but-1-ene by this mechanism, producing two isomeric products.
      (i) Write a mechanism for the reaction of HBr with CH₂=CHCH₂CH₃, showing the formation of the major product only. Name the product.   [3 marks]
      (ii) Explain why it is the major product for this reaction.   [2 marks]

This section is free from all GM ingredients...

Wow... these pages really are jam-packed. There's not one, not two, but three mechanisms to learn. And learn them you must. They mightn't be as handy in real life as a tin opener, but you won't need a tin opener in the exam. Get the book shut and scribble them out. Make sure your arrows start at the electron pair and finish exactly where the electrons are going.

E/Z Isomers and Polymers

The chemistry on these pages isn't so bad. And don't be too worried when I tell you that a good working knowledge of German would be useful. It's not absolutely essential... and you'll be fine without.

Double Bonds Can't Rotate

1) Carbon atoms in a C=C double bond and the atoms bonded to these carbons all lie in the **same plane** (they're **planar**).
Because of the way they're arranged, they're actually said to be **trigonal planar** — the atoms attached to each double-bond carbon are at the corners of an imaginary equilateral triangle.

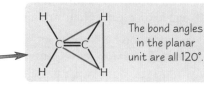

The bond angles in the planar unit are all 120°.

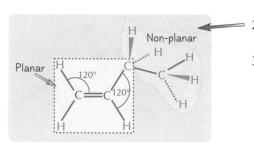

2) Ethene, C_2H_4 (like in the diagram above) is completely planar, but in larger alkenes, only the >C=C< unit is planar.

3) Another important thing about C=C double bonds is that atoms **can't rotate** around them like they can around single bonds. In fact, double bonds are fairly **rigid** — they don't bend much either.

4) Even though atoms can't rotate about the **double bond**, things can still rotate about any **single bonds** in the molecule — like in this molecule of pent-2-ene.

5) The **restricted rotation** around the C=C double bond is what causes **E/Z isomerism**.

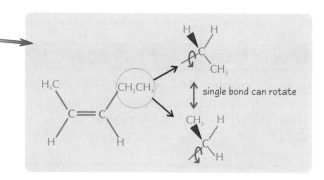

E/Z isomerism is a Type of Stereoisomerism

1) **Stereoisomers** have the same structural formula but a **different arrangement** in space.
(Just bear with me for a moment... that will become clearer, I promise.)

2) Because of the **lack of rotation** around the double bond, some **alkenes** can have stereoisomers.

3) Stereoisomers happen when the two double-bonded carbon atoms each have **different atoms** or **groups** attached to them. Then you get an '**E-isomer**' and a '**Z-isomer**'.

For example, the double-bonded carbon atoms in but-2-ene each have an **H** and a CH_3 group attached.

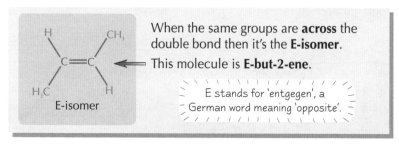

When the same groups are **across** the double bond then it's the **E-isomer**.
This molecule is **E-but-2-ene**.

E stands for 'entgegen', a German word meaning 'opposite'.

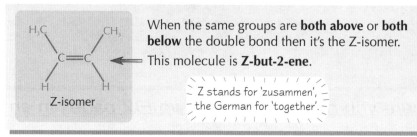

When the same groups are **both above** or **both below** the double bond then it's the Z-isomer.
This molecule is **Z-but-2-ene**.

Z stands for 'zusammen', the German for 'together'.

E/Z Isomers and Polymers

Alkenes *Join Up* to form *Addition Polymers*

1) The **double bonds** in alkenes can open up and join together to make long chains called **polymers**. It's kind of like they're holding hands in a big line. The individual, small alkenes are called **monomers**.

2) This is called **addition polymerisation**. For example, **poly(ethene)** is made by the **addition polymerisation** of ethene.

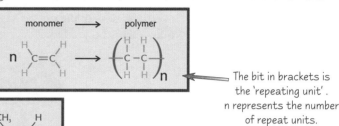

The bit in brackets is the 'repeating unit'. n represents the number of repeat units.

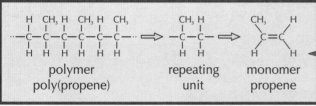

3) To find the **monomer** used to form an addition polymer, take the **repeating unit** and add a **double bond**.

4) Because of the loss of the double bond, poly(alkenes), like alkanes, are **unreactive**.

5) Different polymer **structures** have different **properties**, which means they're suited to different **uses**. Some typical uses of **poly(ethene)** and **poly(propene)** are shown in the table.

6) Many polymers are difficult to dispose of and are made from **non-renewable** oil fractions, so it makes sense to **recycle** them. For example, poly(propene) is recycled — it can be **melted** and **remoulded**.

	Properties	Uses
Low density poly(ethene)	Soft Flexible	Plastic bags Squeezy bottles
Poly(propene)	Tough Strong	Bottle crates Rope

Practice Questions

Q1 Why is an ethene molecule said to be planar?

Q2 What feature of alkene molecules gives rise to E/Z isomerism?

Q3 Which of the following is the Z-isomer of but-2-ene?

Q4 Define the term 'stereoisomers'.

Q5 Give a typical use for poly(ethene), and one for poly(propene).

Exam Questions

Q1 a) Draw and name the E/Z isomers of pent-2-ene. [4 marks]
    b) Explain why alkenes can have E/Z isomers but alkanes cannot. [2 marks]

Q2 Part of the structure of a polymer is shown on the right.
    a) (i) Draw the repeating unit of the polymer. [1 mark]
       (ii) Draw the monomer from which the polymer was formed. [1 mark]

    b) Poly(tetrafluoroethene) is made from the monomer shown on the right. Draw part of the polymer consisting of three of the repeating units. [1 mark]

And there you have it, folks — two E/Z pages in an AS Chemistry book...

To remember which is which out of E and Z isomers, remember that Z-isomers are the ones with the groups on 'ze zame zide'. Or if you prefer, you could learn to speak German... It's the double bond in alkenes that gives them their special powers — things can't rotate around it, so there are stereoisomers. And it can open up, so they can form addition polymers.

Alcohols

Alcohol — evil stuff, it is. I could start preaching, but I won't, because this page is enough to put you off alcohol for life...

Alcohols can be **Primary**, **Secondary** or **Tertiary**

1) The alcohol homologous series has the **general formula $C_nH_{2n+1}OH$**.

2) An alcohol is **primary**, **secondary** or **tertiary**, depending on which carbon atom the hydroxyl group **–OH** is bonded to...

> The naming rules on page 36-37 apply to alcohols too. The suffix for alcohols is –ol, and you have to say which carbon the –OH group is attached to.

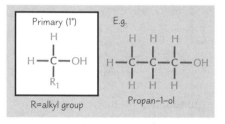

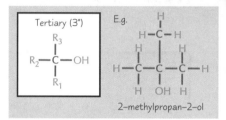

Ethanol can be Produced Industrially by **Fermentation**

At the moment most industrial ethanol is produced by **steam hydration of ethene** with a **phosphoric acid catalyst** (see page 73). The ethene comes from cracking heavy fractions of crude oil. But in the future, when crude oil supplies start **running out**, petrochemicals like ethene will be expensive — so producing ethanol by **fermentation** will become much more important...

Industrial Production of Ethanol by Fermentation

1) Fermentation is an **exothermic** process, carried out by **yeast** in **anaerobic conditions** (without oxygen).

2) Yeast produces an **enzyme** which converts sugars, such as glucose, into **ethanol** and **carbon dioxide**.

3) The enzyme works at an **optimum** (ideal) temperature of **30-40 °C**. If it's too cold, the reaction is **slow** — if it's too hot, the enzyme is **denatured** (damaged).

4) When the solution reaches about **15% ethanol**, the yeast dies. **Fractional distillation** is used to increase the concentration of ethanol.

5) Fermentation is **low-tech** — it uses cheap equipment and **renewable resources**. The ethanol produced by this method has to be **purified** though.

$$C_6H_{12}O_{6(aq)} \xrightarrow[\text{yeast}]{30\text{-}40°C} 2C_2H_5OH_{(aq)} + 2CO_{2(g)}$$
glucose

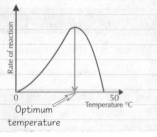

Learn the **Pros and Cons** of Each Method of Making **Ethanol**

Here's a quick summary of the **advantages** and **disadvantages** of the two methods of making ethanol.

Method	Rate of reaction	Quality of product	Raw material	Process/Costs
Hydration of ethene	Very fast	Pure	Ethene from oil — a finite resource	Continuous process, so expensive equipment needed, but low labour costs.
Fermentation	Very slow	Very impure — needs further processing	Sugars — a renewable resource	Batch process, so cheap equipment needed, but high labour costs.

Ethanol is an **Almost Carbon Neutral** Fuel

1) Ethanol is also being used increasingly as a **fuel**, particularly in countries with few oil reserves. E.g. in Brazil, **sugars** from sugar cane are **fermented** to produce alcohol, which is added to petrol. Ethanol, made in this way, is a **biofuel**.

> A biofuel is a fuel that's made from biological material that's recently died.

2) Ethanol is thought of as a **carbon neutral** fuel, because all the CO_2 released when the fuel is burned was removed by the crop as it grew. **BUT** — there are still carbon emissions if you consider the **whole** process. E.g. making the fertilisers and powering agricultural machinery will probably involve burning fossil fuels.

Alcohols

Alcohols can be **Dehydrated** to Form **Alkenes**

1) You can make ethene by **eliminating** water from **ethanol** in a **dehydration reaction** (i.e. elimination of **water**).

$$C_2H_5OH \longrightarrow CH_2{=}CH_2 + H_2O$$

In an elimination reaction, a small group of atoms breaks away from a larger molecule. It's not replaced by anything else.

Here's how you can go about it:

Dehydrating Alcohols to form Alkenes

You have to **reflux** ethanol with **concentrated sulfuric acid**.

Check out the next page for what refluxing is.

The reaction occurs in two stages:

This is the reverse of the hydrolysis reaction on p73.

$$C_2H_5OH + H_2SO_4 \longrightarrow C_2H_5OSO_2OH + H_2O$$
$$C_2H_5OSO_2OH \longrightarrow CH_2{=}CH_2 + H_2SO_4$$

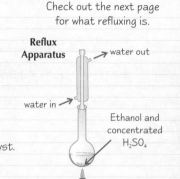

Reflux Apparatus
water out
water in
Ethanol and concentrated H_2SO_4
heat

The H_2SO_4 is unchanged at the end of the reaction, so it's acted as a catalyst. Phosphoric acid (H_3PO_4) can also be used as a catalyst for this reaction.

2) This reaction allows you to produce alkenes from **renewable** resources.

Because you can produce ethanol by fermentation of glucose, which you can get from plants.

3) This is important, because it means that you can produce **polymers** (poly(ethene), for example) **without** needing **oil**.

Practice Questions

Q1 What is the general formula for an alcohol?

Q2 Write down two advantages and two disadvantages of the fermentation method of producing ethanol.

Q3 What is a biofuel? What does carbon-neutral mean? Is ethanol a carbon-neutral biofuel?

Q4 Describe how to form an alkene from an alcohol using an acid catalysed elimination reaction.

Exam Questions

Q1 Butanol C_4H_9OH has four chain and positional isomers.
Name each isomer and class it as a primary, secondary or tertiary alcohol.

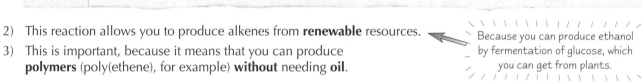

a)
```
  H  H  H  H
  |  |  |  |
H-C- C - C - C-OH
  |  |  |  |
  H  H  H  H
```

b)
```
      H
    H-C-H
  H   |   H
  |   |   |
H-C- C - C - H
  |   |   |
  H  OH  H
```

c)
```
  H  H  H  H
  |  |  |  |
H-C- C - C - C-H
  |  |  |  |
  H  H  OH H
```

d)
```
      OH
    H-C-H
  H   |   H
  |   |   |
H-C- C - C - H
  |   |   |
  H  H   H
```

[8 marks]

Q2 Ethanol is a useful alcohol.
   a) State whether ethanol is a primary, secondary or tertiary alcohol, and explain why. [2 marks]
   b) Industrially, ethanol can be produced by fermentation of glucose, $C_6H_{12}O_6$.
      (i) Write a balanced equation for this reaction. [1 mark]
      (ii) State the optimum conditions for fermentation. [3 marks]
   c) At present most ethanol is produced by the acid-catalysed hydration of ethene.
      Why is this? Why might this change in the future? [3 marks]

Euuurghh, what a page... I think I need a drink...

Not much to learn here — a few basic definitions, an industrial process, the advantages and disadvantages of it compared to another industrial process, a bit about biofuels, and a lovely two-stage reaction... Like I said, not much here at all. Think I'm going to faint. [THWACK]

Oxidising Alcohols

Another page of alcohol reactions. Probably not what you wanted for Christmas... But at least you are almost at the end of the section... and the book for that matter... and your wits, probably.

How Much an Alcohol can be **Oxidised** Depends on its **Structure**

The simple way to oxidise alcohols is to burn them. But you don't get the most exciting products by doing this. If you want to end up with something more interesting at the end, you need a more sophisticated way of oxidising...

You can use the **oxidising agent acidified potassium dichromate(VI)** to **mildly** oxidise alcohols.

- **Primary** alcohols are oxidised to **aldehydes** and then to **carboxylic acids**.
- **Secondary** alcohols are oxidised to **ketones** only.
- **Tertiary** alcohols aren't oxidised.

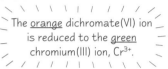 The <u>orange</u> dichromate(VI) ion is reduced to the <u>green</u> chromium(III) ion, Cr^{3+}.

Learn What **Aldehydes**, **Ketones** and **Carboxylic Acids** are

Aldehydes and **ketones** are **carbonyl** compounds — they have the functional group C=O. Their general formula is $C_nH_{2n}O$.

Aldehydes have a **hydrogen** and **one alkyl group** attached to the carbonyl carbon atom. E.g. Their suffix is **-al**. You don't have to say which carbon the functional group is on — it's always on carbon-1.

propanal
CH_3CH_2CHO

Ketones have **two alkyl groups** attached to the carbonyl carbon atom.

Their suffix is **-one**. For ketones with five or more carbons, you always have to say which carbon the functional group is on. (If there are other groups attached, such as methyl groups, you have to say it for four-carbon ketones too.)

propanone
CH_3COCH_3

pentan-2-one
$CH_3COC_3H_7$

Carboxylic acids have a different functional group...

Carboxylic acids have a **COOH** group at the end of their carbon chain. Their suffix is **-oic**.

propanoic acid CH_3CH_2COOH

Primary Alcohols will Oxidise to *Aldehydes* and *Carboxylic Acids*

$$R-CH_2-OH + [O] \longrightarrow R-C\overset{O}{\underset{H}{\lozenge}} + [O] \xrightarrow{reflux} R-C\overset{O}{\underset{OH}{\lozenge}}$$

[O] = oxidising agent

+ H_2O

primary alcohol   aldehyde   carboxylic acid

You can control how **far** the alcohol is oxidised by controlling the **reaction conditions**:

Oxidising Primary Alcohols

1) Gently heating ethanol with potassium dichromate(VI) solution and sulfuric acid in a test tube should produce "apple" smelling **ethanal** (an aldehyde). However, it's **really tricky** to control the amount of heat and the aldehyde is usually oxidised to form "vinegar" smelling **ethanoic acid**.

2) To get just the **aldehyde**, you need to get it out of the oxidising solution **as soon** as it's formed. You can do this by gently heating excess alcohol with a **controlled** amount of oxidising agent in **distillation apparatus**, so the aldehyde (which boils at a lower temperature than the alcohol) is distilled off **immediately**.

Reflux Apparatus
water out
Liebig condenser
water in
round bottomed flask
anti-bumping granules (added to make boiling smoother)
heat

3) To produce the **carboxylic acid**, the alcohol has to be **vigorously oxidised**. The alcohol is mixed with excess oxidising agent and heated under **reflux**. Heating under reflux means you can increase the **temperature** of an organic reaction to boiling without losing **volatile** solvents, reactants or products. Any vaporised compounds are cooled, condense and drip back into the reaction mixture. Handy, hey.

Oxidising Alcohols

Secondary Alcohols will Oxidise to Ketones

1) Refluxing a secondary alcohol, e.g. propan-2-ol, with acidified dichromate(VI) will produce a **ketone**.
2) Ketones can't be oxidised easily, so even prolonged refluxing won't produce anything more.

Tertiary Alcohols can't be Oxidised Easily

Tertiary alcohols don't react with potassium dichromate(VI) at all — the solution stays orange. The only way to oxidise tertiary alcohols is by **burning** them.

Use Oxidising Agents to Distinguish Between Aldehydes and Ketones

Aldehydes and ketones can be distinguished using **oxidising agents** — aldehydes are easily oxidised but ketones aren't.

1) **Fehling's solution** and **Benedict's solution** are both deep blue Cu^{2+} complexes, which reduce to brick-red Cu_2O when warmed with an aldehyde, but stay blue with a ketone.
2) **Tollen's reagent** is $[Ag(NH_3)_2]^+$ — it's reduced to **silver** when warmed with an aldehyde, but not with a ketone. The silver will coat the inside of the apparatus to form a **silver mirror**.

Practice Questions

Q1 What's the difference between the structure of an aldehyde and a ketone?

Q2 What will acidified potassium dichromate(VI) oxidise secondary alcohols to?

Q3 What is the colour change when potassium dichromate(VI) is reduced?

Q4 Describe two tests you can use to distinguish between a sample of an aldehyde and a sample of a ketone.

Exam Question

Q1 A student wanted to produce the aldehyde propanal from propanol, and set up a reflux apparatus using a suitable oxidising agent.

a) (i) Suggest an oxidising agent that the student could use. [1 mark]
(ii) Draw the structural formula of propanal. [1 mark]

b) The student tested his product and found that he had not produced propanal.
(i) Describe a test for an aldehyde. [2 marks]
(ii) What is the student's product? [1 mark]
(iii) Write equations to show the two-stage reaction. Use [O] to represent the oxidising agent. [2 marks]
(iv) What technique should the student have used and why? [2 marks]

c) The student also tried to oxidise 2-methylpropan-2-ol, unsuccessfully.
(i) Draw the full structural formula for 2-methylpropan-2-ol. [1 mark]
(ii) Why is it not possible to oxidise 2-methylpropan-2-ol with an oxidising agent? [1 mark]

I.... I .... just can't do it, R2...

Don't give up now. Only as a fully-trained Chemistry Jedi, with the force as your ally, can you take on the Examiner. If you quit now, if you choose the easy path as Wader did, all the marks you've fought for will be lost. Be strong. Don't give in to hate — that leads to the dark side... (Only two more double pages to go before you're done with this section...)

Analytical Techniques

Get ready for the thrilling climax of the book — and watch out for the twist at the end...

Mass Spectrometry Can Help to Identify Compounds

1) You saw on pages 8-9 how **mass spectrometry** can be used to find **relative isotopic masses**, the **abundance** of different isotopes, and the **relative molecular mass**, M_r, of a compound.

2) Remember — to find the relative molecular mass of a compound you look at the **molecular ion peak** (the **M peak**) on the spectrum. Molecular ions are formed when molecules have **electrons** knocked off. The mass/charge value of the molecular ion peak is the molecular mass.

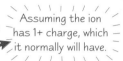
Assuming the ion has 1+ charge, which it normally will have.

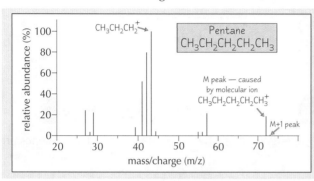

Here's the mass spectrum of pentane. Its M peak is at 72 — so the compound's M_r is 72.

For most <u>organic compounds</u> the M peak is the one with the second highest mass/charge ratio. The smaller peak to the right of the M peak is called the M+1 peak — it's caused by the presence of the carbon isotope ^{13}C (you don't need to worry about this at AS).

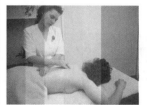

A massage spectrum

The Molecular Ion can be Broken into Smaller Fragments

The bombarding electrons make some of the molecular ions break up into **fragments**. The fragments that are ions show up on the mass spectrum, making a **fragmentation pattern**. Fragmentation patterns are actually pretty cool because you can use them to identify **molecules** and even their **structure**.

For propane, the molecular ion is $CH_3CH_2CH_3^+$, and the fragments it breaks into include CH_3^+ ($M_r = 15$) and $CH_3CH_2^+$ ($M_r = 29$).

Only the **ions** show up on the mass spectrum — the **free radicals** are 'lost'.

$$CH_3CH_2CH_3^+ \begin{cases} CH_3CH_2\bullet + CH_3^+ \\ \text{free radical} \quad \text{ion} \\ \\ CH_3CH_2^+ + \bullet CH_3 \\ \text{ion} \quad \text{free radical} \end{cases}$$

To work out the structural formula, you've got to work out what **ion** could have made each peak from its **m/z value**. (You assume that the m/z value of a peak matches the **mass** of the ion that made it.)

Example: Use this mass spectrum to work out the structure of the molecule:

It's only the m/z values you're interested in — ignore the heights of the bars.

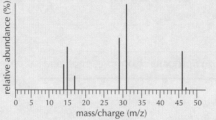

Fragment	Molecular Mass
CH_3	15
C_2H_5	29
C_3H_7	43
OH	17

1. Identify the fragments

This molecule's got a peak at 15 m/z, so it's likely to have a **CH₃ group**.

It's also got a peak at 17 m/z, so it's likely to have an **OH group**.

Other ions are matched to the peaks here:

2. Piece them together to form a molecule with the correct M_r

Ethanol has all the fragments on this spectrum.

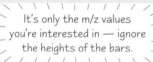

Ethanol's **molecular mass** is 46. This should be the same as the m/z value of the M peak — it is.

Analytical Techniques

If you've got some stuff and don't know what it is, don't taste it. Stick it in an infrared spectrometer instead. Infrared spectroscopy produces scary looking graphs. But just learn the basics, and you'll be fine.

Infrared Spectroscopy Helps You Identify Organic Molecules

1) In infrared (IR) spectroscopy, a beam of **IR radiation** is passed through a sample of a chemical.

2) The IR radiation is absorbed by the **covalent bonds** in the molecules, increasing their **vibrational** energy.

3) **Bonds between different atoms** absorb **different frequencies** of IR radiation. Bonds in different **places** in a molecule absorb different frequencies too — so the O–H group in an **alcohol** and the O–H in a **carboxylic acid** absorb different frequencies. This table shows what **frequencies** different bonds absorb:

Functional group	Where it's found	Frequency/ Wavenumber (cm^{-1})	Type of absorption
C–H	most organic molecules	2800 - 3100	strong, sharp
O–H	alcohols	3200 - 3550	strong, broad
O–H	carboxylic acids	2500 - 3300	medium, broad
N–H	amines (e.g. methylamine, CH_3NH_2)	3200 - 3500	strong, sharp
C=O	aldehydes, ketones, carboxylic acids	1680 - 1750	strong, sharp
C–X	haloalkanes	500 - 1000	strong, sharp

This tells you what the peak on the graph will look like.

You don't need to learn this data, but you do need to understand how to use it.

4) An infrared spectrometer produces a **graph** that shows you what frequencies of radiation the molecules are absorbing. So you can use it to identify the **functional groups** in a molecule:

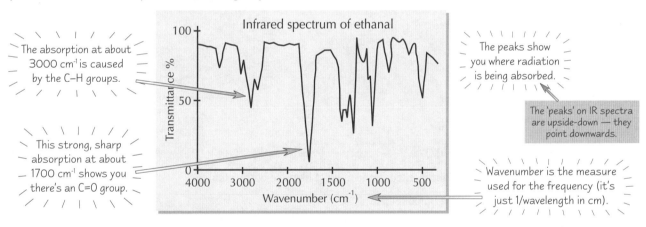

The absorption at about 3000 cm^{-1} is caused by the C–H groups.

This strong, sharp absorption at about 1700 cm^{-1} shows you there's an C=O group.

The peaks show you where radiation is being absorbed.

The 'peaks' on IR spectra are upside-down — they point downwards.

Wavenumber is the measure used for the frequency (it's just 1/wavelength in cm).

The Fingerprint Region Identifies a Molecule

1) The region between **1000 cm^{-1} and 1550 cm^{-1}** on the spectrum is called the **fingerprint** region. It's **unique** to a **particular compound**. You can check this region of an unknown compound's IR spectrum against those of known compounds. If it **matches up** with one of them, hey presto — you know what the molecule is.

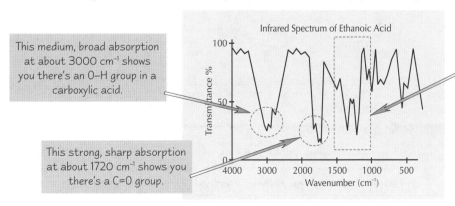

This medium, broad absorption at about 3000 cm^{-1} shows you there's an O–H group in a carboxylic acid.

This strong, sharp absorption at about 1720 cm^{-1} shows you there's a C=O group.

This is the fingerprint region. If you see an infrared spectrum of an unknown molecule that has the same pattern in this area, you can be sure that it's ethanoic acid.

Clark began to regret having an infrared mechanism installed in his glasses.

2) Infrared spectroscopy can also be used to find out how **pure** a compound is, and identify any impurities. Impurities produce **extra peaks** in the fingerprint region.

Analytical Techniques

Infrared Energy *Absorption* is Linked to *Global Warming*

1) Some of the electromagnetic radiation emitted by the **Sun** is in the form of **infrared radiation**.

2) Molecules of **greenhouse gases**, like **carbon dioxide**, **methane** and **water vapour**, are really good at absorbing infrared energy — so if the amounts of them in the atmosphere increase, it leads to **global warming**. There's lots more about how this happens on pages 42-43.

It's the bonds of these molecules that absorb the IR radiation.

Practice Questions

Q1 What is meant by the molecular ion?

Q2 What is the M peak?

Q3 How do fragments get formed?

Q4 Which parts of a molecule absorb infrared energy?

Q5 Why do most infrared spectra of organic molecules have a strong, sharp peak at around 3000 cm⁻¹?

Q6 On an infrared spectrum, what is meant by the 'fingerprint region'?

Exam Question

Q1 On the right is the mass spectrum of an organic compound, Q.

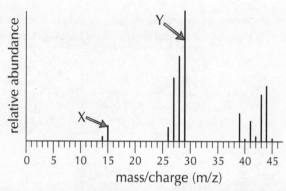

a) What is the M_r of compound Q? [1 mark]

b) What fragments are the peaks marked X and Y most likely to correspond to? [2 marks]

c) Suggest a structure for this compound. [1 mark]

d) Why is it unlikely that this compound is an alcohol? [2 marks]

Q2 A molecule with a molecular mass of 74 produces the following IR spectrum.

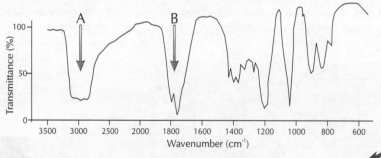

Use the infrared absorption data on the previous page.

a) Which functional groups are responsible for peaks A and B? [2 marks]

b) Suggest the molecular formula and name of this molecule. Explain your answer. [3 marks]

I wonder what the infrared spectrum of a fairy cake would look like...

I don't suppose I'll ever know. Very squiggly I imagine. Luckily you don't have to be able to remember what any infrared spectrum graphs look like. But you definitely need to know how to interpret them. And don't worry, I haven't forgotten I said there was twist at the end... erm... hydrogen was my sister all along... and all the elements went to live in Jamaica. The End.

Practical and Investigative Skills

You're going to have to do some practical work too — and once you've done it, you have to make sense of your results...

Make it a **Fair Test** — Control your **Variables**

You probably know this all off by heart but it's easy to get mixed up sometimes. So here's a quick recap:

> **Variable** — A variable is a **quantity** that has the **potential to change**, e.g. mass.
> There are two types of variable commonly referred to in experiments:
> - **Independent variable** — the thing that you **change** in an experiment.
> - **Dependent variable** — the thing that you **measure** in an experiment.

When drawing graphs, the dependent variable should go on the y-axis, the independent on the x-axis.

So, if you're investigating the effect of **temperature** on rate of reaction using the apparatus on the right, the variables will be:

Independent variable	Temperature
Dependent variable	Amount of oxygen produced — you can measure this by collecting it in a gas syringe
Other variables — you MUST keep these the same	Concentration and volume of solutions, mass of solids, pressure, the presence of a catalyst and the surface area of any solid reactants

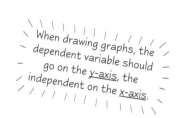

Know Your Different Sorts of **Data**

Experiments always involve some sort of measurement to provide **data**.
There are different types of data — and you need to know what they are.

> **Discrete** — you get discrete data by **counting**. E.g. the number of bubbles produced in a reaction would be discrete. You can't have 1.25 bubbles. That'd be daft. Shoe size is another good example of a discrete variable.

> **Continuous** — a continuous variable can have **any value** on a scale. For example, the volume of gas produced or the mass of products from a reaction. You can never measure the exact value of a continuous variable.

> **Categoric** — a categoric variable has values that can be sorted into **categories**. For example, the colours of solutions might be blue, red and green. Or types of material might be wood, steel, glass.

> **Ordered (ordinal)** — Ordered data is similar to categoric, but the categories can be **put in order**. For example, if you classify reactions as 'slow', 'fairly fast' and 'very fast' you'd have ordered data.

Organise Your Results in a **Table** — And Watch Out For **Anomalous** Ones

Before you start your experiment, make a **table** to write your results in.
You'll need to repeat each test at least three times to check your results are reliable.

This is the sort of table you might end up with when you investigate the effect of **temperature** on **reaction rate**.
(You'd then have to do the same for **different temperatures**.)

Temperature	Time (s)	Volume of gas evolved (cm³) Run 1	Volume of gas evolved (cm³) Run 2	Volume of gas evolved (cm³) Run 3	Average volume of gas evolved (cm³)
	10	8	7	8	7.7
20 °C	20	17	19	20	18.7
	30	28	(20)	30	29

Find the average of each set of repeated values.

You need to add them all up and divide by how many there are.

E.g.: (8 + 7 + 8) ÷ 3 = 7.7 cm³

Watch out for **anomalous results**. These are ones that don't fit in with the other values and are likely to be wrong. They're likely to be due to random errors — here the syringe plunger may have got stuck.
Ignore anomalous results when you calculate the average.

Practical and Investigative Skills

Graphs: *Line, Bar or Scatter — Use the Best Type*

You'll usually be expected to make a **graph** of your results. Not only are graphs **pretty**, they make your data **easier to understand** — so long as you choose the right type.

Line graphs are best when you have **two sets of continuous data**. For example:

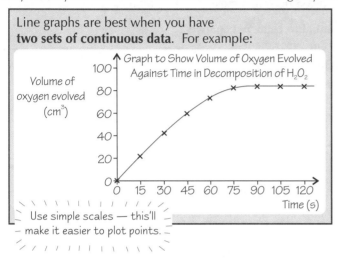

Use simple scales — this'll make it easier to plot points.

You should use a bar chart when one of your data sets is **categoric or ordered data**. For example:

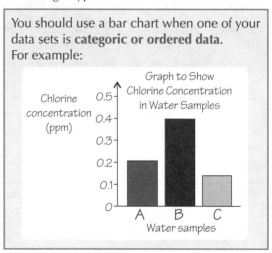

Scatter plots are great for showing how two sets of data are related (or **correlated**).

Don't try to join all the points — draw a **line of best fit** to show the **trend**.

Scatter Graph to Show Relationship Between Relative Molecular Masses and Melting Points of Straight-Chain Alcohols

Scatter Graphs Show The Relationship Between Variables

Correlation describes the **relationship** between two variables — the independent one and the dependent one.

Data can show:

1) **Positive correlation** — as one variable **increases** the other **increases**. The graph on the left shows positive correlation.

2) **Negative correlation** — as one variable **increases** the other **decreases**.

3) **No correlation** — there is **no relationship** between the two variables.

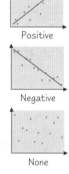

There are also pie charts. These are normally used to display categoric data.

Whatever type of graph you make, you'll ONLY get full marks if you:

• Choose a sensible scale — don't do a tiny graph in the corner of the paper.

• Label both axes — including units.

• Plot your points accurately — using a sharp pencil.

Correlation *Doesn't* Mean *Cause* — Don't Jump to Conclusions

1) Ideally, only **two** quantities would **ever** change in any experiment — everything else would remain **constant**.

2) But in experiments or studies outside the lab, you **can't** usually control all the variables. So even if two variables are correlated, the change in one may **not** be causing the change in the other. Both changes might be caused be a **third variable**.

Watch out for bias too — for instance, a bottled water company might point these studies out to people without mentioning any of the doubts.

Example

For example: Some studies have found a correlation between **drinking chlorinated tap water** and the risk of developing certain cancers. So some people argue that this means water shouldn't have chlorine added.

BUT it's hard to control all the variables between people who drink tap water and people who don't. It could be many lifestyle factors.

Or, the cancer risk could be affected by something else in tap water — or by whatever the non-tap water drinkers drink instead...

Practical and Investigative Skills

Don't Get Carried Away When Drawing Conclusions

The **data** should always **support** the conclusion. This may sound obvious but it's easy to **jump** to conclusions. Conclusions have to be **specific** — not make sweeping generalisations.

> **Example**
>
> The rate of an enzyme-controlled reaction was measured at **10 °C, 20 °C, 30 °C, 40 °C, 50 °C and 60 °C**. All other variables were kept constant, and the results are shown in this graph.
>
> A science magazine **concluded** from this data that enzyme X works best at **40 °C**. The data **doesn't** support this.
>
> The enzyme **could** work best at 42 °C or 47 °C but you can't tell from the data because **increases** of **10 °C** at a time were used. The rate of reaction at in-between temperatures **wasn't** measured.
>
> All you know is that it's faster at **40 °C** than at any of the other temperatures tested.

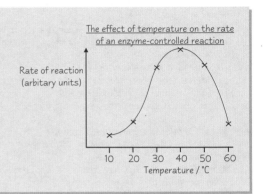

The effect of temperature on the rate of an enzyme-controlled reaction

Rate of reaction (arbitary units)

Temperature / °C

> **Example**
>
> The experiment above **ONLY** gives information about this particular enzyme-controlled reaction. You can't conclude that **all** enzyme-controlled reactions happen faster at a particular temperature — only this one. And you can't say for sure that doing the experiment at, say, a different constant pressure, wouldn't give a different optimum temperature.

You need to Look Critically at Your Results

There are a few bits of lingo that you need to understand. They'll be useful when you're evaluating how convincing your results are.

1) **Valid results** — Valid results answer the original question. For example, if you haven't **controlled all the variables** your results won't be valid, because you won't be testing just the thing you wanted to.

2) **Accurate** — Accurate results are those that are **really close** to the **true** answer.

3) **Precise results** — These are results taken using **sensitive instruments** that measure in **small increments**, e.g. pH measured with a meter (pH 7.692) will be **more precise** than pH measured with paper (pH 7).

   > It's possible for results to be precise **but not** accurate, e.g. a balance that weighs to 1/1000 th of a gram will give precise results but if it's not **calibrated** properly the results won't be accurate.

   You may have to calculate the percentage error of a measurement.
   E.g. if a balance is calibrated to within 0.1 g, and you measure a mass as 4 g, then the percentage error is: $(0.1 \div 4) \times 100 = 2.5\%$.
   Using a larger quantity reduces the percentage error. E.g. a mass of 40 g has a percentage error of: $(0.1 \div 40) \times 100 = 0.25\%$.

4) **Reliable results** — **Reliable** means the results can be **consistently reproduced** in independent experiments. And if the results are reproducible they're more likely to be **true**. If the data isn't reliable for whatever reason you **can't draw** a valid **conclusion**.

   For experiments, the **more repeats** you do, the **more reliable** the data. If you get the **same result** twice, it could be the correct answer. But if you get the same result **20 times**, it'd be much more reliable. And it'd be even more reliable if everyone in the class got about the same results using different apparatus.

Work Safely and Ethically — Don't Blow Up the Lab or Harm Small Animals

In any experiment you'll be expected to show that you've thought about the **risks and hazards**. It's generally a good thing to wear an apron and goggles, but you may need to take additional safety measures, depending on the experiment. For example, anything involving nasty gases will need to be done in a fume cupboard.

You need to make sure you're working **ethically** too. This is most important if there are other people or animals involved. You have to put their welfare first.

Answers

Unit 1: Section 1 — Atomic Structure

Page 5 — The Atom

1) a) Similarity — They've all got the same number of protons/electrons. [1 mark]
   Difference — They all have different numbers of neutrons. [1 mark]
   b) 1 proton [1 mark], 1 neutron (2 – 1) [1 mark], 1 electron [1 mark].
   c) ^3H. [1 mark]
   Since tritium has 2 neutrons in the nucleus and also 1 proton, it has a mass number of 3. You could also write ^3_1H but you don't really need the atomic number.

2) a) (i) Same number of electrons. [1 mark]
   $^{32}S^{2-}$ has 16 + 2 = 18 electrons. ^{40}Ar has 18 electrons too. [1 mark]
   (ii) Same number of protons. [1 mark].
   Each has 16 protons (the atomic number of S must always be the same) [1 mark].
   (iii) Same number of neutrons [1 mark]
   ^{40}Ar has 40 – 18 = 22 neutrons. ^{42}Ca has 42 – 20 = 22 neutrons. [1 mark]
   b) **A** and **C**. [1 mark] They have the same number of protons but different numbers of neutrons. [1 mark].
   It doesn't matter that they have a different number of electrons because they are still the same element.

Page 7 — Atomic Models

1 a) In order to explain the observations made during his students' experiments [1 mark].
   b) The Bohr model gives a better explanation of the observations [1 mark] of the frequencies of radiation emitted by atoms [1 mark].
   c) The more accurate models are very complicated / The Bohr model is still useful for explaining most observations [1 mark].

2 a) Ca^{2+} is the more stable [1 mark].
   b) Atomic models predict that noble gas electronic configurations/full electron shells are most stable [1 mark]. Since Ca^{2+} has a noble gas electronic configuration / full electron shells it should be more stable than Ca^+ which does not [1 mark].

Page 9 — Relative Mass

1) a) First multiply each relative abundance by the relative mass —
   $120.8 \times 63 = 7610.4$, $54.0 \times 65 = 3510.0$
   Next add up the products —
   $7610.4 + 3510.0 = 11\,120.4$ [1 mark]
   Now divide by the total abundance (120.8 + 54.0 = 174.8)

   $$A_r(Cu) = \frac{11120.4}{174.8} \approx \textbf{63.6}$$ [1 mark]

   You can check your answer by seeing if A_r(Cu) is in between 63 and 65 (the lowest and highest relative isotopic masses).
   b) A sample of copper is a mixture of 2 isotopes in different abundances [1 mark]. The weighted average mass of these isotopes isn't a whole number [1 mark].

2) a) Mass spectrometry. [1 mark]
   b) You use pretty much the same method here as for question 1)a).
   $93.11 \times 39 = 3631.29$, $(0.12 \times 40) = 4.8$, $(6.77 \times 41) = 277.57$
   $3631.29 + 4.8 + 277.57 = 3913.66$ [1 mark]
   This time you divide by 100 because they're percentages.

   $$A_r(K) = \frac{3913.66}{100} \approx \textbf{39.14}$$ [1 mark]

   Again check your answer's between the lowest and highest relative isotopic masses, 39 and 41. A_r(K) is closer to 39 because most of the sample (93.11 %) is made up of this isotope.

Page 11 — Electronic Structure

1) a) K atom: $1s^2\,2s^2\,2p^6\,3s^2\,3p^6\,4s^1$ [1 mark]
   K^+ ion: $1s^2\,2s^2\,2p^6\,3s^2\,3p^6$ [1 mark]
   b)
   Oxygen electron Configuration

	1s	2s	2p		
	↑↓	↑↓	↑↓	↑	↑

   1 mark for the correct number of electrons in each sub-shell.
   1 mark for having spin-pairing in one of the p orbitals and parallel spins in the other two p orbitals. A box filled with 2 arrows is spin pairing — 1 up and 1 down. If you've put the four p electrons into just 2 orbitals, it's wrong.

c) The outer shell electrons in potassium and oxygen can get close to the outer shells of other atoms so they can be transferred or shared [1 mark]. The inner shell electrons are tightly held and shielded from the electrons in other atoms/molecules [1 mark].

2) a) $1s^2\,2s^2\,2p^6\,3s^2\,3p^6\,3d^5\,4s^2$. [1 mark]
   b)
   Al^{3+} electron Configuration

	1s	2s	2p		
	↑↓	↑↓	↑↓	↑↓	↑↓

   1 mark for the correct number of electrons in each sub-shell.
   1 mark for one arrow in each box pointing up, and one pointing down.
   c) Germanium ($1s^2\,2s^2\,2p^6\,3s^2\,3p^6\,3d^{10}\,4s^2\,4p^2$). [1 mark].
   (The 4p sub-shell is partly filled so it must be a p block element.)
   d) Ar (atom) [1 mark], K^+ (positive ion) [1 mark], Cl^- (negative ion) [1 mark]. You also could have suggested Ca^{2+}, S^{2-} or P^{3-}.

Page 13 — Ionisation Energies

1) a) $Li_{(g)} \rightarrow Li^+_{(g)} + e^-$ [1 mark for the correct equation with wrong or missing state symbols. 1 mark for the correct state symbols.]
   b) Increasing number of protons means a stronger pull from the positively charged nucleus [1 mark] making it harder to remove an electron from the outer shell [1 mark]. There are no extra inner electrons to add to the shielding effect [1 mark].
   c) (i) Boron has the configuration $1s^2\,2s^2\,2p^1$ compared to $1s^2\,2s^2$ for beryllium [1 mark]. The 2p shell is at a slightly higher energy level than the 2s shell. As a result, the extra distance and partial shielding of the 2s orbital make it easier to remove the outer electron [1 mark].
   (ii) Oxygen has the configuration $1s^2\,2s^2\,2p^4$ compared to $1s^2\,2s^2\,2p^3$ for nitrogen [1 mark]. Electron repulsion in the shared 2p sub-shell in oxygen makes it easier to remove an electron [1 mark].

2) As you go down Group 2, it takes less energy to remove an electron [1 mark]. This is evidence that the outer electrons are increasingly distant from the nucleus [1 mark] and additional inner shells of electrons exist to shield the outer shell [1 mark].

Unit 1: Section 2 — Amount of Substance

Page 15 — The Mole

1) M of $CH_3COOH = (2 \times 12) + (4 \times 1) + (2 \times 16) = 60$ g mol^{-1} [1 mark]
   so mass of 0.36 moles $= 60 \times 0.36 = \textbf{21.6 g}$ [1 mark]

2) No. of moles $= \frac{0.25 \times 60}{1000} = 0.015$ moles H_2SO_4 [1 mark]
   M of $H_2SO_4 = (2 \times 1) + (1 \times 32) + (4 \times 16) = 98$ g mol^{-1}
   Mass of 0.015 mol $H_2SO_4 = 98 \times 0.015 = \textbf{1.47 g}$ [1 mark]

3) a) M of $C_3H_8 = (3 \times 12) + (8 \times 1) = 44$ g mol^{-1}
   No. of moles of $C_3H_8 = \frac{88}{44} = 2$ moles [1 mark]
   At r.t.p. 1 mole of gas occupies 24 dm^3
   so 2 moles of gas occupies $2 \times 24 = \textbf{48 dm}^3$ [1 mark]
   b) $pV = nRT$ [1 mark]
   so $V = nRT \div p = [2 \times 8.31 \times 308] \div [100 \times 10^3]$
   $= 51.1896 \times 10^{-3}$ m^3 $\approx \textbf{51.2 dm}^3$ [1 mark]

Page 17 — Equations and Calculations

1) M of $C_2H_5Cl = (2 \times 12) + (5 \times 1) + (1 \times 35.5) = 64.5$ g mol^{-1} [1 mark]
   Number of moles of $C_2H_5Cl = \frac{258}{64.5} = 4$ moles [1 mark]
   From the equation, 1 mole C_2H_5Cl is made from 1 mole C_2H_4
   so, 4 moles C_2H_5Cl is made from 4 moles C_2H_4 [1 mark].
   M of $C_2H_4 = (2 \times 12) + (4 \times 1) = 28$ g mol^{-1}
   so, the mass of 4 moles $C_2H_4 = 4 \times 28 = \textbf{112 g}$ [1 mark]

2) a) M of $CaCO_3 = 40 + 12 + (3 \times 16) = 100$ g mol^{-1}
   Number of moles of $CaCO_3 = \frac{15}{100} = 0.15$ moles [1 mark]
   From the equation, 1 mole $CaCO_3$ produces 1 mole CaO
   so, 0.15 moles of $CaCO_3$ produces 0.15 moles of CaO [1 mark]
   M of CaO $= 40 + 16 = 56$ g mol^{-1}
   so, mass of 0.15 moles of CaO $= 56 \times 0.15 = \textbf{8.4 g}$ [1 mark]

b) From the equation, 1 mole $CaCO_3$ produces 1 mole CO_2
so, 0.15 moles of $CaCO_3$ produces 0.15 moles of CO_2 [1 mark]
1 mole gas occupies 24 dm³ [1 mark],
so, 0.15 moles occupies = 24 × 0.15 = **3.6 dm³** [1 mark]

3) On the LHS, you need 2 each of K and I, so use 2KI
This makes the final equation: **2KI + Pb(NO₃)₂ → PbI₂ + 2KNO₃**
[1 mark]

In this equation, the NO_3 group remains unchanged, so it makes balancing much easier if you treat it as one indivisible lump.

Page 19 — Titrations

1) First write down what you know —
$CH_3COOH + NaOH \rightarrow CH_3COONa + H_2O$
$\quad$ 25.4 cm³ $\quad\quad$ 14.6 cm³
$\quad\quad$? $\quad\quad\quad$ 0.5 M

Number of moles of NaOH = $\dfrac{0.5 \times 14.6}{1000}$ = 0.0073 moles [1 mark]

From the equation, you know 1 mole NaOH neutralises 1 mole of CH_3COOH, so if you've used 0.0073 moles NaOH you must have neutralised 0.0073 moles CH_3COOH. [1 mark]

Concentration of $CH_3COOH = \dfrac{0.0073 \times 1000}{25.4}$ = **0.287 M** [1 mark]

2) First write down what you know again —
$CaCO_3 + H_2SO_4 \rightarrow CaSO_4 + H_2O + CO_2$
0.75 g $\quad$ 0.25 M

M of $CaCO_3$ = 40 + 12 + (3 × 16) = 100 g mol⁻¹ [1 mark]

Number of moles of $CaCO_3 = \dfrac{0.75}{100}$ = 7.5 x 10⁻³ moles [1 mark]

From the equation, 1 mole $CaCO_3$ reacts with 1 mole H_2SO_4
so, 7.5 × 10⁻³ moles $CaCO_3$ reacts with 7.5 × 10⁻³ moles H_2SO_4. [1 mark]

The volume needed is = $\dfrac{(7.5 \times 10^{-3}) \times 1000}{0.25}$ = 30 cm³ [1 mark]

If the question mentions concentration or molarities, you can bet your last clean pair of underwear that you'll need to use the formula

no. of moles = $\dfrac{\text{conc.} \times \text{volume (cm}^3)}{1000}$ (or no. moles = conc. × volume (dm³)).

Page 21 — Formulas, Yield and Atom Economy

1) Assume you've got 100 g of the compound so you can turn the % straight into mass.

No. of moles of C = $\dfrac{92.3}{12}$ = 7.69 moles

No. of moles of H = $\dfrac{7.7}{1}$ = 7.7 moles [1 mark]

Divide both by the smallest number, in this case 7.69.
So ratio C : H = 1 : 1
So, the empirical formula = CH [1 mark]

The empirical mass = 12 + 1 = 13

No. of empirical units in molecule = $\dfrac{78}{13}$ = 6

So the molecular formula = **C₆H₆** [1 mark]

2)a) There is only one product, so the theoretical yield can be calculated by adding the masses of the reactants [1 mark].
So theoretical yield = 0.275 + 0.142 = 0.417 g [1 mark]
b) percentage yield = (0.198 ÷ 0.417) × 100 = 47.5% [1 mark]
c) Changing reaction conditions will have no effect on atom economy [1 mark]. Since the equation shows that there is only one product, the atom economy will always be 100% [1 mark].
Atom economy is related to the type of reaction — addition, substitution, etc. — not to the quantities of products and reactants.

Unit 1: Section 3 — Bonding and Periodicity

Page 23 — Ionic Bonding

1)a)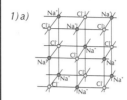

Your diagram should show the following —
• cubic structure with ions at corners [1 mark]
• sodium ions and chloride ions labelled [1 mark]
• alternating sodium ions and chloride ions [1 mark]
b) giant ionic/crystal (lattice) [1 mark]
c) You'd expect it to have a high melting point [1 mark] because there are strong bonds between the ions [1 mark] due to the electrostatic forces [1 mark]. A lot of energy is required to overcome these bonds [1 mark].
2)a) Electrons move from one atom to another [1 mark].
Any correct examples of ions, one positive (e.g. Na⁺) [1 mark], one negative (e.g. Cl⁻) [1 mark].
b) In a solid, ions are held in place by strong ionic bonds [1 mark]. When the solid is heated to melting point, the ions gain enough energy to overcome the forces of attraction enough to become mobile [1 mark] and so carry charge (and hence electricity) through the substance [1 mark].

Page 25 — Covalent Bonding

1)a) Covalent [1 mark]
b)

Your diagram should show the following —
• a completely correct electron arrangement in carbon [1 mark]
• all 4 overlaps correct (one dot and one cross in each) [1 mark]
2)a) Dative covalent/coordinate bonding [1 mark]
b) One atom donates [1 mark] a pair of electrons to the bond [1 mark].
3)a) Giant molecular/macromolecular/giant covalent [1 mark]
b) Diamond Graphite [1 mark for each correctly drawn diagram]

Diamond's a bit awkward to draw without it looking like a bunch of ballet dancing spiders — just make sure each central carbon is connected to four others.
c) Diamond has electrons in localised covalent bonds [1 mark], so is a poor electrical conductor [1 mark]. Graphite has delocalised electrons which can flow within the sheets [1 mark], making it a good electrical conductor [1 mark].

Answers

Page 27 — Shapes of Molecules

1) a) NCl₃   *[1 mark]*   BCl₃   *[1 mark]*

b) NCl₃   ××   *[1 mark]*

shape: (trigonal) pyramidal *[1 mark]*,
bond angle: 107° (accept between 105° and 109°) *[1 mark]*

BCl₃   Cl   Cl   *[1 mark]*

(It must be a reasonable "Y" shaped molecule.)
shape: trigonal planar *[1 mark]*, bond angle: 120° exactly *[1 mark]*

c) BCl₃ has three electron pairs only around B *[1 mark]*.
NCl₃ has four electron pairs around N *[1 mark]*, including one lone pair *[1 mark]*.

Page 30 — Polarisation and Intermolecular Forces

1) a) The power of an atom to withdraw electron density *[1 mark]* from a covalent bond *[1 mark]* OR the ability of an atom to attract the bonding electrons *[1 mark]* in a covalent bond *[1 mark]*.

b) (i)
Br —— Br

(ii)
O δ⁻
δ+ H   H δ+

(iii)
δ⁻ N
H δ+   H   H δ+

[1 mark for each correct shape, 1 mark for bond polarities correctly marked on H₂O, 1 mark for bond polarities correctly marked on NH₃.]

2) a) Van der Waals OR instantaneous/temporary dipole-induced dipole OR dispersion forces.
Permanent dipole-dipole interactions/forces.
Hydrogen bonding.
(Permanent dipole-induced dipole interactions.)
[1 mark each for any three]

b)

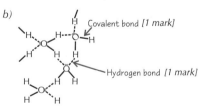

Covalent bond *[1 mark]*
Hydrogen bond *[1 mark]*

You could have shown the H₂O molecules in either of these two ways.

    [1 mark]

Van der Waals OR instantaneous/temporary dipole-induced dipole OR dispersion forces of attraction between water molecules. *[1 mark]*

c) More energy *[1 mark]* is needed to break the hydrogen bonds between water molecules *[1 mark]*.

Page 33 — Metallic Bonding and Properties of Structures

1)

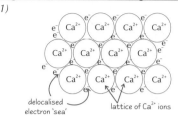

delocalised electron 'sea'
lattice of Ca²⁺ ions

[1 mark for showing closely packed Ca²⁺ ions and 1 mark for showing a sea of delocalised electrons.]
Metallic bonding results from the attraction between positive metal ions *[1 mark]* and a sea of delocalised electrons between them *[1 mark]*.

2) a) A — Ionic   B — (Simple) molecular
C — Metallic   D — Giant molecular (macromolecular)
[1 mark for each]

b) (i) Diamond — D   (ii) Aluminium — C
(iii) Sodium chloride — A (iv) Iodine — B
[2 marks if all correct. 1 mark for only two correct.]

3) **Magnesium** has a metallic crystal lattice (it has metallic bonding) *[1 mark]*. It has a sea of electrons/delocalised electrons/freely moving electrons *[1 mark]*, which allow it to conduct electricity in the solid or liquid state *[1 mark]*.
Sodium chloride has a (giant) ionic lattice *[1 mark]*. It doesn't conduct electricity when it's solid *[1 mark]* because its ions don't move freely, but vibrate about a fixed point *[1 mark]*. Sodium chloride conducts electricity when liquid/molten *[1 mark]* or in aqueous solution *[1 mark]* because it has freely moving ions (not electrons) *[1 mark]*.
Graphite is giant covalent/macromolecular *[1 mark]*. It has delocalised/freely moving electrons between the layers *[1 mark]*. It conducts electricity along the layers in the solid state *[1 mark]*.

Page 35 — Periodicity

1) Mg has more delocalised electrons per atom *[1 mark]* and the ion has a greater charge density (due to its smaller ionic radius) *[1 mark]*. This gives Mg a stronger metal-metal bond, resulting in a higher boiling point *[1 mark]*.

2) a) Si has a macromolecular (or giant molecular) structure *[1 mark]* consisting of very strong covalent bonds *[1 mark]*.

b) Sulfur (S₈) has a larger molecule than phosphorus (P₄) *[1 mark]*. which results in larger van der Waals forces of attraction between molecules *[1 mark]*.

3) The atomic radius decreases across the period from left to right *[1 mark]*. The number of protons increases, so nuclear charge increases *[1 mark]*. Electrons are pulled closer to the nucleus *[1 mark]*. The electrons are all added to the same outer shell, so there's little effect on shielding *[1 mark]*.

4) Neon has the configuration $1s^2 2s^2 2p^6$ and sodium $1s^2 2s^2 2p^6 3s^1$. *[1 mark]* The extra distance of sodium's outer electron from the nucleus and electron shielding make it easier to remove an electron from the 3s sub-shell *[1 mark]*.

Unit 1: Section 4 — Alkanes and Organic Chemistry

Page 37 — Basic Stuff

1) a)

1-bromobutane

[1 mark]

b) Haloalkanes / halogenoalkanes *[1 mark]*

c) but-1-ene *[2 marks, or 1 mark for just butene]*

2) a) 1-chloro-2-methylpropane
[2 marks available, lose 1 mark for each mistake]
Remember to put the substituents in alphabetical order.

b) 3-methylbut-1-ene *[2 marks available, lose 1 mark for each mistake]*

c) 2,4-dibromo-but-1-ene
[2 marks available, lose 1 mark for each mistake]
In parts b) and c), the double bond is the most important functional group, so it's given the lowest number.

Answers

Page 39 — Formulas and Structural Isomerism

1) a) 1-chlorobutane, 2-chlorobutane, 1-chloro-2-methylpropane, 2-chloro-2-methylpropane [1 mark for each correct isomer]

b) 1-chloro-2-methylpropane and 2-chloro-2-methylpropane
OR 1-chlorobutane and 2-chlorobutane [1 mark]

c) 1-chlorobutane and 1-chloro-2-methylpropane
OR 2-chlorobutane and 2-chloro-2-methylpropane [1 mark]

2) a)

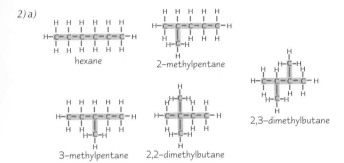

hexane   2–methylpentane

3–methylpentane   2,2–dimethylbutane   2,3–dimethylbutane

[1 mark for each correctly drawn isomer,
1 mark for each correct name]

b) A group of compounds represented by the same general formula OR having the same functional group OR with similar chemical properties [1 mark]. Each successive member differs by $-CH_2-$ [1 mark].

c) (i) C_8H_{18} [1 mark]
(ii) $CH_3CH_2CH_2CH_2CH_2CH_2CH_2CH_3$ [1 mark]

3) a) pentane, (2-)methylbutane, (2,2-)dimethylpropane [1 mark for each]
There's only actually one type of methylbutane. You can't have 1-methylbutane — it'd be exactly the same as pentane.

b) Chain isomers have the same molecular formula/are made up of the same atoms [1 mark], but they have different arrangements of the carbon skeleton [1 mark].

Page 41 — Alkanes and Petroleum

1) a) As a mixture, crude oil is not very useful — the different alkanes it's made up of have different uses [1 mark].

b) Boiling point [1 mark].

c) (i) C_8H_{18} [1 mark]
(ii) Near the top [1 mark]. This is because the molecules in petrol have a relatively low boiling point [1 mark] and the fractionating column is cooler at the top than the bottom [1 mark].
(iii) C_8H_{16} [1 mark]

2) a) There's greater demand for smaller fractions [1 mark] for motor fuels [1 mark] OR for alkenes [1 mark] to make petrochemicals/ polymers [1 mark].

b) E.g. $C_{12}H_{26} \rightarrow C_2H_4 + C_{10}H_{22}$ [1 mark].
There are loads of possible answers — just make sure the C's and H's balance and there's an alkane and an alkene.

Page 43 — Alkanes as Fuels

1) a) $C_7H_{16} + 11O_2 \rightarrow 7CO_2 + 8H_2O$
[2 marks available, lose 1 mark for each error]

b) (i) Carbon monoxide [1 mark]
(ii) By fitting a catalytic converter [1 mark]

2 a) Nitrogen and oxygen from air [1 mark] react together because of the conditions (high pressure and temperature) in the engine [1 mark].

b) Some fossil fuels contain sulfur that forms sulfur dioxide when burned [1 mark]. Sulfur dioxide dissolves in water in the atmosphere to form an acid (sulfuric acid) [1 mark]. Power stations remove the sulfur dioxide from their flue gases using calcium oxide [1 mark].

Unit 2: Section 1 — Energetics

Page 45 — Enthalpy Changes

1) a) Total energy required to break bonds = (4 × 435) + (2 × 498)
= 2736 kJ mol⁻¹ [1 mark]
Energy released when bonds form = (2 × 805) + (4 × 464)
= 3466 kJ mol⁻¹ [1 mark]
Net energy change = +2736 + (−3466) = −730 kJ mol⁻¹
[1 mark for correct numerical value, 1 mark for correct unit]

b) The reaction is exothermic, because the enthalpy change is negative / more energy is given out than is taken in [1 mark].

2) a) $CH_3OH_{(l)} + 1\frac{1}{2}O_{2(g)} \rightarrow CO_{2(g)} + 2H_2O_{(l)}$
Correct balanced equation [1 mark]. Correct state symbols for reactants [1 mark].
It is perfectly OK to use halves to balance equations. Make sure that only 1 mole of CH_3OH is combusted, as it says in the definition for $\Delta H_c^{\ominus}$.

b) $C_{(s)} + 2H_{2(g)} + \frac{1}{2}O_{2(g)} \rightarrow CH_3OH_{(l)}$
Correct balanced equation [1 mark]. Correct state symbols for reactants [1 mark].

c) Only 1 mole of C_3H_8 should be shown according to the definition of $\Delta H_c^{\ominus}$ [1 mark].
You really need to know the definitions of the standard enthalpy changes off by heart. There's loads of nit-picky little details they could ask you questions about.

Page 47 — Calculating Enthalpy Changes

1) $\Delta H_r^{\ominus}$ = sum of $\Delta H_f^{\ominus}$(products) − sum of $\Delta H_f^{\ominus}$(reactants) [1 mark]
= [0 + (3 × −602)] − [−1676 + (3 × 0)] [1 mark]
= −130 kJ mol⁻¹ [1 mark]
Don't forget the units. It's a daft way to lose marks.

2) No. of moles of $CuSO_4 = \frac{0.200 \times 50}{1000}$ [1 mark] = 0.01 moles [1 mark]
From the equation, 1 mole of $CuSO_4$ reacts with 1 mole of Zn.
So, 0.01 moles of $CuSO_4$ reacts with 0.01 moles of Zn [1 mark].
Heat produced by reaction = mc∆T [1 mark]
= 50 × 4.18 × 2.6 = 543.4 J [1 mark]
0.01 moles of zinc produces 543.4 J of heat, therefore 1 mole of zinc
produces $\frac{543.4}{0.01}$ [1 mark] = 54 340 J ≈ 54.3 kJ
So the enthalpy change is −54.3 kJ mol⁻¹ (you need the minus sign because it's exothermic) [1 mark for correct number, 1 mark for minus sign].
It'd be dead easy to work out the heat produced by the reactions, breathe a sigh of relief and sail on to the next question. But you need to find out the enthalpy change when 1 mole of zinc reacts. It's always a good idea to reread the question and check you've actually answered it.

Unit 2: Section 2 — Kinetics and Equilibria

Page 50 — Reaction Rates and Catalysts

1) The molecules don't always have enough energy [1 mark].

2) The particles in a liquid move freely and all of them are able to collide with the solid particles [1 mark]. Particles in solids just vibrate about fixed positions, so only those on the touching surfaces between the two solids will be able to react. [1 mark]

3) a) $2H_2O_{2(l)} \rightarrow 2H_2O_{(l)} + O_{2(g)}$
Correct symbols [1 mark] and balancing equation [1 mark]. You get the marks even if you forgot the state symbols.

b)

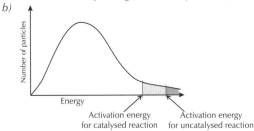

Activation energy for catalysed reaction   Activation energy for uncatalysed reaction

Correct general shape of the curve [1 mark]. Correctly labelling the axes [1 mark]. Activation energies marked on the horizontal axis — the catalysed activation energy must be lower than the uncatalysed activation energy [1 mark].

Answers

c) Manganese(IV) oxide lowers the activation energy by providing an alternative reaction pathway [1 mark]. So, more reactant molecules have at least the activation energy [1 mark], meaning there are more successful collisions in a given period of time, and so the rate increases [1 mark].

d) Raising the temperature will increase the rate of reaction [1 mark]. The particles will have more kinetic energy on average [1 mark], so a greater proportion of particles will have enough energy to react [1 mark].

Page 53 — Reversible Reactions

1) a) If a reaction at equilibrium is subjected to a change in concentration, pressure or temperature, the equilibrium will shift to try to oppose (counteract) the change. [1 mark].
Examiners are always asking for definitions so learn them — they're easy marks.

b) (i) There's no change [1 mark]. There's the same number of molecules/moles on each side of the equation [1 mark].

(ii) Reducing temperature removes heat. So the equilibrium shifts in the exothermic direction to release heat [1 mark]. The reverse reaction is exothermic (since the forward reaction is endothermic). So, the position of equilibrium shifts left [1 mark].

(iii) Removing nitrogen monoxide reduces its concentration. The equilibrium position shifts right to try and increase the nitrogen monoxide concentration again [1 mark].

c) No effect [1 mark].
Catalysts don't affect the equilibrium position.
They just help the reaction to get there sooner.

2) a) At low temperature the particles move more slowly / have less energy [1 mark]. This means fewer successful collisions [1 mark] and a slower rate of reaction [1 mark].

b) High pressure is expensive. [1 mark] The cost of the extra pressure is greater than the value of the extra yield. [1 mark]

Unit 2: Section 3 — Reactions and Elements

Page 55 — Redox Reactions

1) a) Oxidation is the loss of electrons, reduction is the gain of electrons [1 mark].

b) (i) 0 [1 mark]   (ii) +1 [1 mark]

c) $Li \rightarrow Li^+ + e^-$
[2 marks — 1 mark for electrons on RHS, 1 mark for correct equation, allow different balancing]
$O_2 + 4e^- \rightarrow 2O^{2-}$
[2 marks — 1 mark for electrons on LHS, 1 mark for correct equation, allow different balancing]
Lithium is being oxidised and oxygen is being reduced [1 mark].

2) a) An oxidising agent accepts electrons from another species [1 mark].

b) $In \rightarrow In^{3+} + 3e^-$
[2 marks — 1 mark for electrons on RHS, 1 mark for correct equation]

c) $2In + 3Cl_2 \rightarrow 2InCl_3$ [2 marks — 1 mark for correct reactants and product, 1 mark for correct balancing]

Page 57 — Group 7 — The Halogens

1) a) $I_2 + 2At^- \rightarrow 2I^- + At_2$ [1 mark]

b) The (sodium) astatide [1 mark]

2) a) (i) Boiling point increases down the group [1 mark] because the size and relative mass of the atoms increases [1 mark], so the Van der Waals forces holding the molecules together get stronger [1 mark].

(ii) Electronegativity decreases down the group [1 mark] because the atoms get larger [1 mark], and larger atoms don't attract electrons as strongly as smaller ones [1 mark].

b) Fluorine [1 mark]

3 a) $Cl_2 + H_2O \rightarrow HCl$ [1 mark] $+ HClO$ [1 mark]

b) Chlorine (or the chlorate(I) ions) kill bacteria [1 mark].
Too much chlorine would be dangerous because it is toxic [1 mark].

Page 59 — Halide Ions

1) **Aqueous** solutions of both halides are tested.   [1 mark]

a) **Sodium chloride** — silver nitrate gives white precipitate which dissolves in dilute ammonia solution [1 mark].
$Ag^+ + Cl^- \rightarrow AgCl$ [1 mark]
Sodium bromide — silver nitrate gives cream precipitate which is only soluble in concentrated ammonia solution [1 mark].
$Ag^+ + Br^- \rightarrow AgBr$ [1 mark]

b) **Sodium chloride** — Misty fumes [1 mark]
$NaCl + H_2SO_4 \rightarrow NaHSO_4 + HCl$ [1 mark]
Sodium bromide — Misty fumes [1 mark]
$NaBr + H_2SO_4 \rightarrow NaHSO_4 + HBr$ [1 mark]
$2HBr + H_2SO_4 \rightarrow Br_2 + SO_2 + 2H_2O$ [1 mark]
Orange / brown vapour [1 mark]

2) a) NaI (via HI) reduces H_2SO_4 to H_2S [1 mark]. The reducing power of halide ions increases down the group [1 mark] and At is below I in the group [1 mark], so H_2S will be produced [1 mark].

b) AgI is insoluble in concentrated ammonia solution [1 mark]. The solubility of halides in ammonia solution decreases down the group [1 mark], so AgAt will **NOT** dissolve. [1 mark]
Question 2 is the kind of question that could completely throw you if you're not really clued up on the facts. If you really know p58, then in part a) you'll go, "Ah - ha!!! Reactions of halides with H_2SO_4 — reducing power increases down the group..." If not, you basically won't have a clue. The moral is... it really is just about learning all the facts. Boring, but true.

Page 62 — Group 2 — The Alkaline Earth Metals

1) $Mg \quad 1s^2\ 2s^2\ 2p^6\ 3s^2 \qquad Ca \quad 1s^2\ 2s^2 2p^6 3s^2 3p^6\ 4s^2$ [1 mark]
First ionisation energy of Ca is smaller [1 mark] because Ca has (one) more electron shell(s) [1 mark]. This reduces the attraction between the nucleus and the outer electrons because it increases shielding the effect [1 mark] and because the outer shell of Ca is further from the nucleus [1 mark].

2) a) Y [1 mark]

b) Y has the largest radius [1 mark] so it will have the smallest ionisation energy/lose its outer electrons more easily [1 mark].

3) Add barium chloride (or nitrate) solution to both [1 mark]
Zinc chloride would not change/no reaction [1 mark]
Zinc sulfate solution would give a white precipitate [1 mark]
$BaCl_{2(aq)} + ZnSO_{4(aq)} \rightarrow BaSO_{4(s)} + ZnCl_{2(aq)}$ [1 mark]
(or suitable equation using $Ba(NO_3)_{2(aq)}$)
OR $Ba^{2+}_{(aq)} + SO_4^{2-}_{(aq)} \rightarrow BaSO_{4(s)}$ [1 mark]
(or a test with silver nitrate for the chloride ions could be done.)

4) a) $NaHCO_3 + HCl \rightarrow NaCl + H_2O + CO_2$ [1 mark]

b) Wind/burping etc. [1 mark]

c) Magnesium hydroxide [1 mark] – NB many other of the compounds would be either toxic or otherwise harmful.
$Mg(OH)_2 + 2HCl \rightarrow MgCl_2 + 2H_2O$ [1 mark]

Page 65 — Extraction of Metals

1) a) $Fe_3O_4 + 4CO \rightarrow 3Fe + 4CO_2$ [1 mark]

b) $Fe_3O_4 + 2C \rightarrow 3Fe + 2CO_2$
OR $Fe_3O_4 + 4C \rightarrow 3Fe + 4CO$ [1 mark]

2) Advantage: the tungsten produced is purer [1 mark]
Disadvantage: hydrogen is more expensive
OR hydrogen is highly explosive [1 mark]

3) a) Aluminium oxide dissolved [1 mark] in molten cryolite [1 mark]

b) Cathode: $Al^{3+} + 3e^- \rightarrow Al$ [1 mark]
Anode: $2O^{2-} \rightarrow O_2 + 4e^-$ [1 mark]

c) High energy costs of extracting Al [1 mark].

Unit 2: Section 4 — More Organic Chemistry

Page 67 — Synthesis of Chloroalkanes

1) a) One with no double bonds OR the maximum number of hydrogens OR single bonds only [1 mark]. It contains only hydrogen and carbon atoms [1 mark].

b) It has non-polar bonds/it's a non-polar molecule [1 mark], so it does not attract/react with polar reagents [1 mark].

Answers

c) $CH_3CH_3 + Cl_2 \xrightarrow{U.V.} CH_3CH_2Cl + HCl$ *[1 mark]*
Initiation: $Cl_2 \xrightarrow{U.V.} 2Cl\cdot$ *[1 mark]*
Propagation: $CH_3CH_3 + Cl\cdot \rightarrow CH_3CH_2\cdot + HCl$ *[1 mark]*
$CH_3CH_2\cdot + Cl_2 \rightarrow CH_3CH_2Cl + Cl\cdot$ *[1 mark]*
Termination: $CH_3CH_2\cdot + Cl\cdot \rightarrow CH_3CH_2Cl$
Or: $CH_3CH_2\cdot + CH_3CH_2\cdot \rightarrow CH_3CH_2CH_2CH_3$ *[1 mark]*
[1 mark for mentioning U.V.]
It's a free-radical [1 mark] substitution [1 mark] reaction.

Page 70 — Nucleophilic Substitution and Elimination
1) a) **Reaction 1**
Reagent — NaOH/KOH/OH⁻ *[1 mark]*
Solvent — Water/aqueous solution *[1 mark]*
Reaction 2
Reagent — Ammonia/NH_3 *[1 mark]*
Solvent — Ethanol/alcohol *[1 mark]*
Reaction 3
Reagent — NaOH/KOH *[1 mark]*.
Solvent — Ethanol/alcohol *[1 mark]*
   b) *There'd be a faster reaction [1 mark]. The C–I bond is weaker than C–Br, or C–I bond enthalpy is lower [1 mark].*

Page 73 — Reactions of Alkenes
1) a) *Shake the alkene with bromine water [1 mark], and the solution goes colourless if a double bond is present [1 mark].*
   b) *Electrophilic [1 mark] addition [1 mark].*
   c) (i)

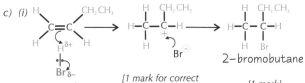

2-bromobutane

[1 mark] *[1 mark for correct intermediate]* *[1 mark]*

Check that your curly arrows are exactly right, or you'll lose marks. They have to go from exactly where the electrons are from, to where they're going to.
   (ii) *The secondary carbocation OR the carbocation with the most attached alkyl groups [1 mark] is the most stable intermediate and so is the most likely to form [1 mark].*

Page 75 — E/Z Isomers and Polymers
1) a)

H₃C, H, CH₂CH₃ *[1 mark]*
E-pent-2-ene *[1 mark]*

H₃C, CH₂CH₃, H *[1 mark]*
Z-pent-2-ene *[1 mark]*

   b) *E/Z isomers occur because atoms can't rotate about C=C double bonds [1 mark]. Alkenes contain C=C double bonds and alkanes don't, so alkenes can form E/Z isomers and alkanes can't [1 mark].*
2) a) (i) *[1 mark]*

   (ii) *[1 mark]*

   b) *[1 mark]*

Page 77 — Alcohols
1) a) *Butan-1-ol [1 mark], primary [1 mark]*
   b) *2-methylpropan-2-ol [1 mark], tertiary [1 mark]*
   c) *Butan-2-ol [1 mark], secondary [1 mark]*
   d) *2-methylpropan-1-ol [1 mark], primary [1 mark]*
2) a) *Primary [1 mark]. The -OH group is bonded to a carbon with one alkyl group/other carbon atom attached [1 mark].*
   b) (i) $C_6H_{12}O_{6(aq)} \rightarrow 2C_2H_5OH_{(aq)} + 2CO_{2(g)}$ *[1 mark]*
      (ii) *Yeast [1 mark], temperature between 30 and 40 °C [1 mark], Anaerobic conditions OR air/oxygen excluded [1 mark]*
   c) *Ethene is cheap and abundantly available / It's a low-cost process / it's a high-yield process / Very pure ethanol is produced / Fast reaction [1 mark each for up to two of these reasons]. This might change in the future as crude oil reserves run out / become more expensive [1 mark].*

Page 79 — Oxidising Alcohols
1) a) (i) *Acidified potassium dichromate(VI) [1 mark]*
      (ii)

propanal *[1 mark]*
CH_3CH_2CHO

   b) (i) *Warm with Fehling's/Benedict's solution: turns from blue to brick-red OR warm with Tollen's reagent: a silver mirror is produced [1 mark for test, 1 mark for result]*
      (ii) *Propanoic acid [1 mark]*
      (iii) $CH_3CH_2CH_2OH + [O] \rightarrow CH_3CH_2CHO + H_2O$ *[1 mark]*
$CH_3CH_2CHO + [O] \rightarrow CH_3CH_2COOH$ *[1 mark]*
      (iv) *Distillation [1 mark]. This is so aldehyde is removed immediately as it forms [1 mark].*
If you don't get the aldehyde out quick-smart, it'll be a carboxylic acid before you know it.
   c) (i)

[1 mark]
      (ii) *2-methylpropan-2-ol is a tertiary alcohol (which is more stable) [1 mark].*

Page 82 — Analytical Techniques
1) a) *44 [1 mark]*
   b) *X has a mass of 15. It is probably a methyl group/CH_3. [1 mark] Y has a mass of 29. It is probably an ethyl group/C_2H_5. [1 mark]*
   c) *[1 mark]*

   d) *If the compound was an alcohol, you would expect a peak with m/z ratio of 17 [1 mark], caused by the OH fragment [1 mark].*
2 a) *A's due to an O–H group in a carboxylic acid [1 mark]. B's due to a C=O as in an aldehyde, ketone, acid or ester [1 mark].*
   b) *The spectrum suggests it's a carboxylic acid — it's got a COOH group [1 mark]. This group has a mass of 45, so the rest of the molecule has a mass of 29 (74 – 45), which is likely to be C_2H_5 [1 mark]. So the molecule could be C_2H_5COOH — propanoic acid [1 mark].*

Index

A

absorption (of infrared radiation) 81, 82
accurate results 85
acid rain 42
activation energy 48-50
addition reactions 71-73
addition polymerisation 75
alcohols 68, 70, 73, 76-79
aldehydes 70, 78, 79
alkaline earth metals 60-62
alkanes 36, 40, 42
alkenes 37, 69-73, 75, 77
alkyl groups 72
alpha particles 6
aluminium 64
amines 69
ammonia 25, 26, 59, 69
ammonium ion 25
amount of substance 14
analytical techniques 80, 81
anomalous results 83
aromatic hydrocarbons 41
atom economy 21
atomic (proton) number 4
atomic models 6, 7
atomic number 4, 5
atomic radius 34, 60
atoms 4-7
average bond enthalpies 44, 45
Avogadro's constant 14

B

balancing equations 16
barium chloride 61
barium meals 62
barium sulfate 62
bauxite 64
Benedict's solution 79
biofuel 76
bleach 56
Bohr, Niels 6
boiling points 29, 32, 35, 56
bond angles 26, 27
bond enthalpies 44, 45, 68
bonding 22-25, 31-33
bonding pairs 26
breaking bonds 44, 45
bromine 32, 56
bromine water 71
burettes 18
burning alkanes 42

C

calcium hydroxide 62
calculating
    atom economy 21
    concentrations 14, 18
    empirical formulas 20
    gas volumes 15, 17
    masses 14, 16
    molecular formulas 20
    solution volumes 19
    percentage yield 20
calorimeters 46
carbocations 71, 72
carbon 24, 25, 63
carbon dioxide 42
carbon monoxide 42, 53, 63
carbon neutrality 53, 76
carbon-12 8
carbonyl compounds 78
carboxylic acids 78
catalysts 49-51, 67, 73
    catalytic converters 42
    catalytic cracking 41

CFCs 66, 67
chain isomers 38
charge 4, 16, 22, 28
charge clouds 26
charge density 35
chlorate(I) ions 56, 57
chlorine 32, 56, 57
    chlorine free radicals 66, 67
chloroalkanes 66
chlorofluorocarbons (CFCs)
        66, 67
chloromethane 66
collision theory 48
combustion
        complete 42
        enthalpy of 45
        incomplete 42
compound ions 54
compounds 22
compromise conditions 52
concentration 14, 18, 49, 51, 52
conclusions 3, 85
controlling variables 3, 84
coordinate covalent bonding 25
copper 63, 65
correlation 84
covalent bonding 24, 25, 28, 35
cows 43
cracking 41
crude oil 40
cryolite 64
crystal lattice 23, 25, 32
cyanide 69
cycloalkanes 40

D

ΔH 44, 45
data, types of 83
d block 11, 34
Daisy the cow 62
Dalton, John 6
dative covalent bonding 25
dehydration of alcohols 77
delocalised electrons 24, 31, 32, 35
density 32
dependent variables 83
diamond 25
dipole-dipole forces 28, 29
dipoles 28, 29
displacement reactions 56
displayed formulas 38
disproportionation 56, 57
distance from nucleus 12
distillation 76, 78
donor atom 25
dot-and-cross diagrams 22, 24, 25, 33
double covalent bonds 24, 37, 71, 74
drawing conclusions 3, 85
dynamic equilibrium 51

E

E/Z isomerism 74
electrical conductivity 23-25, 31, 32
electrolysis 64
electron clouds 29
electron configurations 10, 11, 34
electron density 27, 28
electron repulsion 13
electron shells 10, 34
electronegativity 28, 30, 56, 68
electronic structure 10, 11, 34
electrons 4, 6, 7, 12, 13, 16, 22,
        24-26, 29, 34
electrophiles 71
electrophilic addition 71-73

electrostatic attraction 22, 28
elements 4, 22, 34
elimination 69, 70, 77
empirical formulas 20, 38
endothermic reactions 44, 52
energy 44
energy levels 10
enthalpy changes 44-47
enthalpy profile diagrams 48, 50
equations 16, 17
equilibrium 51, 52
esters 70
ethanol 52, 53, 73, 76, 77
ethanolic ammonia 69
ethene 52, 72
ethical problems 3
evidence 6, 12, 13
exothermic reactions 44, 52
expanding the octet 27
experimental evidence 2, 3, 33
extraction of metals 63-65

F

Fehling's solution 79
fermentation 76
fingerprint region 81
first ionisation energy 12
fluorine 28, 56
formation (enthalpy of) 45
formulas 20
fossil fuels 42
fractional distillation 40, 64
fragmentation patterns 9, 80
free-radical substitution 66
fuels 42
functional groups 36, 38, 81

G

gas constant, R 15
gas volumes 15, 17
gases 31
Geiger, Hans 6
general formulas 38, 76
giant covalent structures 24, 25, 32
giant ionic structures 23
giant metallic structures 31
global warming 42, 43, 82
graphite 24, 32
graphs 84
greenhouse gases 42, 43, 53, 82
ground-level ozone 42
Group 2 60-62
Group 7 56, 57
groups of the periodic table 34

H

half-equations 55
halides 56, 58, 59
haloalkanes (halogenoalkanes) 37, 66, 68-70
halogens 56, 57, 68
heat 46
Hess's law 46, 47
homologous series 38, 76
how science works 2, 3
hydration (steam) 73, 76
hydrocarbons 40, 42, 71
hydrochlorofluorocarbons (HCFCs) 67
hydrofluorocarbons (HFCs) 67
hydrogen 24, 64
hydrogen bonding 29, 30, 32
hydrogen halides 30
hydrolysis 68, 73
hypotheses 2

I

ideal gas equation 15
incomplete combustion 42
independent variables 83
indicators 18
indigestion tablets 62
induced dipole-dipole forces 29
infrared (IR) spectroscopy 81
infrared radiation 81, 82
initiation reactions 66
intermolecular forces 28-30
iodine 29, 56
ionic bonding 22, 23, 32
ionic equations 16, 55
ionisation energies 12, 13, 35, 60, 61
ions 4, 22, 23
iron 63, 65
isomers 38, 39, 74
isotopes 5, 8
isotopic abundance 8, 9
IUPAC naming system 36

K

ketones 70, 78, 79
kinetic energy 31, 48, 49

L

lattice 23, 25, 29, 31, 32
Le Chatelier's principle 51, 52
limitations of bonding models 33
lines of best fit 84
lone pairs 26, 27, 30

M

macromolecular structures 24, 25
magnesium hydroxide 62
magnesium oxide 22
manganese 63
Marsden, Ernest 6
mass (relative) 8
mass number 4
mass spectrometry 8, 9, 80
Maxwell-Boltzmann distribution 48-50
mean bond enthalpies 44, 45
mechanisms 69, 71-73
melting points 23-25, 31, 32, 35, 60
metal extraction 63-65
metallic bonding 31, 32, 60
methane 24, 26, 43, 82
methanol 53
models 6, 33, 48
molar mass 14
molecular formulas 20, 38
molecular lattice 29
molecules 24-29
moles 14, 15
monomers 75

N

neutralisation 18, 19, 62
neutrons 4
nitriles 69
nitrogen oxides 42
nomenclature 36
nuclear model 6
nuclear symbols 4
nucleophiles 68, 69
nucleophilic substitution 68-70
nucleus 4, 6

O

OIL RIG 55
orbitals 4, 10, 13
ores 63, 64

organic chemistry 36-43, 66-82
organising results 83
oxidation 54-56
oxidation states 54
oxides, reduction of 63
oxides of nitrogen 42
oxidising agents 54, 78, 79
oxidising alcohols 78, 79
oxygen 24, 54, 67
ozone, ground level 42
ozone layer 67

P

p block 11, 34
paddy fields 43
peer review 2
percentage composition 20
percentage yield 20, 21
periodic table 5, 34
periodicity 34, 35
periods 12, 34
permanent dipole-dipole forces 28
petrol 41
petroleum 40
phenolphthalein 18
photodissociation 66
physical properties 23-25, 29, 31-33
pipettes 18
plum pudding model 6
polar bonds 28, 66
polarisation 28, 29
pollutants 53
poly(ethene) 75
poly(propene) 75
polymers 75, 77
positional isomers 38
potassium dichromate(VI) 78, 79
precipitate 59, 61
precise results 85
predictions 2, 6
pressure 49, 51, 52
principal quantum number 10
propagation reactions 66
protons 4

Q

quantum model 7

R

rates of reaction 48, 49
reaction (enthalpy of) 45
reactivity
    of Group 2 elements 60
    of haloalkanes 68
    of halogens 56
recycling
    metals 65
    reactants 52
redox reactions 54, 55
reducing agents 54, 63, 64
reduction 54, 55, 63, 64
reflux 69, 78
relative mass 4, 8, 14
    relative atomic mass 8, 9
    relative formula mass 8
    relative isotopic mass 8
    relative molecular mass 8, 14, 15, 80
reliable results 18, 83, 85
reversible reactions 51-53
risks 85
Rutherford, Ernest 6
rutile 64

S

s block 11, 34, 60
saturated hydrocarbons 40
scatterplots 84
scrap metal 65
shapes of molecules 26, 27
shells 6, 10, 22, 24
shielding 12, 13, 34, 58, 60
silver mirror test 79
silver nitrate test for halides 59
simple molecular structures 32
slaked lime 62
smog 42
sodium chlorate(I) 56
sodium chloride 22, 23
solubility 23-25, 31, 32, 61
solutions, concentration of 14, 18
solvents 66
sparingly soluble 61
specific heat capacity 46
spin 10
standard conditions 44
standard enthalpy changes
    of combustion 45
    of formation 45
    of reaction 45
state symbols 17
steam hydration 73, 76
stereoisomers 74
structural formulas 38, 80
structural isomers 38, 39
sub-shells 6, 10, 13, 34
sublimation 24, 25
substitution reactions 68-70
sulfate ions 61
sulfide ores 63
sulfur dioxide 42
sulfuric acid 58, 73, 77

T

temporary dipole 29
tentative nature of scientific knowledge 3
termination reactions 66
theoretical yield 20
theories 2, 3
thermal cracking 41
thermal decomposition 44
titanium 64
titrations 18, 19
Tollen's reagent 79
transition metals 11
triple covalent bonds 24
tungsten 64

U

ultraviolet radiation 66, 67
unburnt hydrocarbons 42
unsaturated hydrocarbons 71

V

Valence-Shell Electron-Pair Repulsion Theory 26
valid results 85
van der Waals forces 29, 32, 35, 56
variables 83, 84
volumes of gases 15, 17

W

water 24, 26, 30, 61
water treatment 57
water vapour 43, 82

Y

yeast 76
yield 20, 52, 53

The Periodic Table

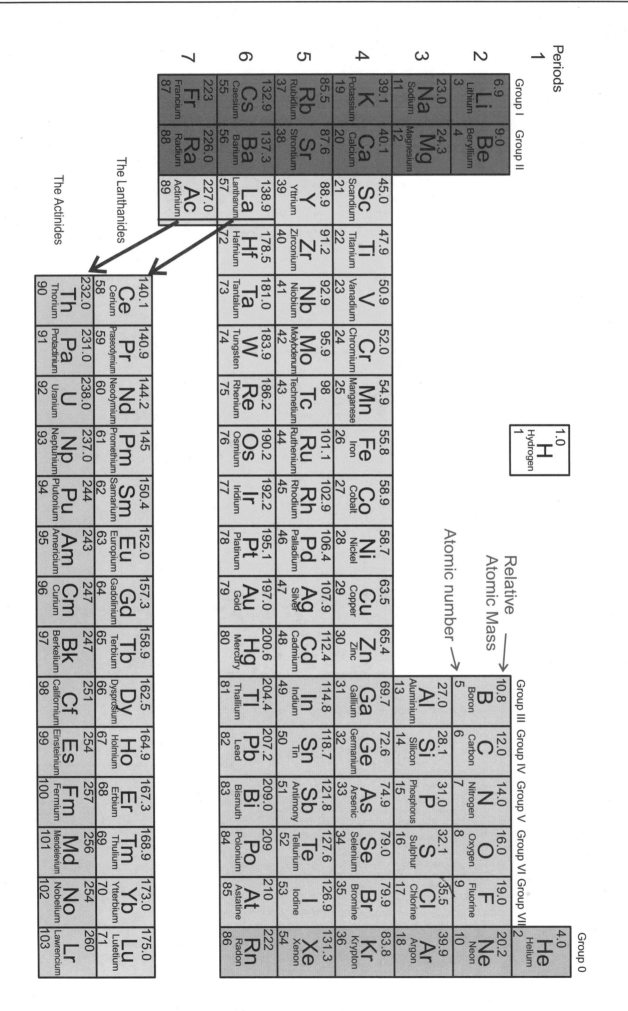

Relative Atomic Mass →

Atomic number →

1.0
H
Hydrogen
1